Collins

KS3 Revision

Maths

Maths

Maths

KS3

Year 8
Workbook

Ian Jacques

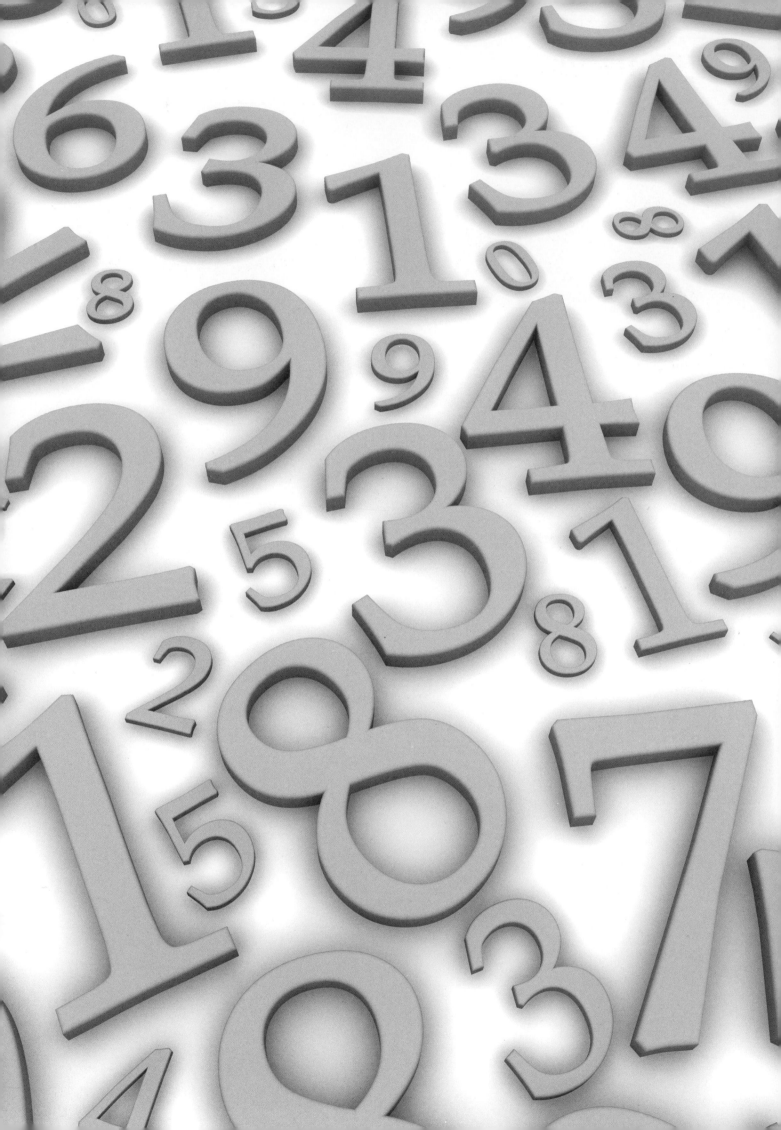

Contents

4 Working with Numbers

8 Geometry

15 Probability

19 Percentages

22 Sequences

26 Area of 2D and 3D Shapes

31 Graphs

36 Simplifying Numbers

38 Interpreting Data

43 Algebra

47 Congruence and Scaling

51 Fractions and Decimals

54 Proportion

57 Circles

61 Equations and Formulae

66 Comparing Data

Answers

1–8 Answers (set in a pull-out booklet in the centre of the book)

Working with Numbers

1 Work out the following:

a) $-4 \times 5 =$ _-20_ ✓

b) $-3 \times -7 =$ _21_ ✓

c) $8 \times -1 =$ _-8_ ✓

d) $-4 \div 2 =$ _-2_ ✓

e) $18 \div -9 =$ _-2_ ✓

f) $-24 \div -6 =$ _4_ ✓

g) $-7 \times 5 =$ _-35_ ✓

h) $-10 \div -5 =$ _-2_ ✗

i) $-4 \times 0 =$ _0_ ✓

j) $(-5)^2 =$ _-25_ ✗

k) $-6 \times 7 =$ _42_ ✓

l) $-100 \div 25 =$ _-4_ ✓　[12]

_10_

2 a) Write down all of the factors of 8.

_1, 2, 4, 8_ ✓　[2] _2_

b) Write down all of the factors of 12.

_1, 2, 3, 4, 6, 12_ ✓　[2] _2_

c) Use your answers to parts a) and b) to write down the common factors of 8 and 12.

_1, 2, 4_ ✓　[1] _1_

d) Use your answer to part c) to write down the highest common factor (HCF) of 8 and 12.

_4_ ✓　[1] _1_

3 a) Write down the first five multiples of 6.

_6/12/18/24/30/36_　[2] _2_

b) Write down the first five multiples of 10.

_10/20/30/40/50/60_　[2] _2_

c) Use your answers to parts a) and b) to write down the lowest common multiple (LCM) of 6 and 10.

_30_　[1] _1_

4 Circle the prime numbers in this list.

②　③　4　⑤　6　⑦　8　9　10　⑪　[2] _2_

5 Use your calculator to work out these powers.

a) $2^7 =$ _128_

b) $3^5 =$ _243_

c) $4^4 =$ _256_　[3] _3_

6 Use your calculator to work out these roots.

a) $\sqrt{441} =$ _21_

b) $\sqrt[3]{216} =$ _6_

c) $\sqrt{72.25} =$ _8.5_　[3] _3_

Total Marks _29_ / 31

1. Fill in the boxes.

a) [3] × −4 = −12

b) −5 × [−6] = −3 × −10

c) [−12]/6 = −2

d) −42/[6] = 14/−2

[4] 4

2. List all of the prime factors of 42.

The first one is 2 then 3 and 7

[2] 2

3. Write down the smallest prime number greater than 50.

53

[2] 2

(PS) 4. Trains leave one of the platforms at a railway station every 15 minutes.
On a different platform trains leave every 20 minutes.

Given that a train leaves both of these platforms at 8:10am what is the next time when these trains will depart both platforms at the same moment?

The answer to this question is 9.10 am.

[3]

5. By using either factor tree or division methods, find the prime factorisation of these numbers.

a) 63

b) 70

$3^2 \times 7$

$2 \times 5 \times 7$

76

[6] 6

6. List all of the factors of the following numbers.

a) 48 ... 1, 3, 3, 4, 5, 6, 12, 16, 24, 48 [2] 2

b) 36 ... 1, 3, 3, 4, 6, 9, 12, 18, 36 [2] 2

c) 18 ... 1, 2, 3, 6, 9, 18 [2] 2

d) 64 ... 1, 2, 4, 8, 16, 32, 64 [2] 2

e) 45 ... 1, 3, 5, 9, 15, 45 [2] 2

7 Use your answers to Q6 to write down the HCF of:

a) 48 and 36   b) 36 and 18   c) 48 and 64   d) 45 and 36

     12   ✓   18   16   9   [4] 4

8 Use your calculator to find all solutions of the following equations.

a) $x^2 = 169$ b) $x^2 = 1.44$ c) $x^2 = 2209$

     13   1.2   47

[6] 6

Total Marks 37 / 37

1 Work out the following.

a) $(2) \times (-1) \times (-3) \times (-4)$

     -24

b) $\dfrac{(3) \times (-1) \times (-4) \times (-2)}{-6}$

     4

c) $\dfrac{(-2)^3 \times (-1)^2}{(-4)}$

[3] 3

(PS) 2 Jonah shares a bedroom with his brothers Bill and Ben.

They all have a cold making it difficult to sleep at night.

Jonah coughs every 8 minutes, Bill coughs every 10 minutes and Ben coughs every 12 minutes.

All three boys cough together at midnight.

a) When do Jonah and Bill next cough at the same time?

     HcF=40. also so it equals 1240cm

[2] 2

b) When do all three boys next cough at the same time?

     HcF 120 so it equals zum.

[2] 2

c) On how many occasions between 1am and 7am do they all cough at the same time?

     3

[1] 1

3 a) Write down two numbers that multiply to 24 and add up to –11.

−8 and −3 [1]

b) Write down two numbers that multiply to –48 and add up to 13.

16 and −3 [1]

(MR) **4** The prime factorisations of two numbers are $2^6 \times 3^4 \times 5^2 \times 7^4$ and $2^5 \times 3^2 \times 5^3 \times 7^4$.
Write down, in index form, the prime factorisation of their:

a) HCF $2^5 \times 3^2 \times 5^2 \times 7^4$ [2]

b) LCM $2^6 \times 3^4 \times 5^3 \times 7^4$ [2]

(MR) **5** a) Find the prime factorisation of 784.

784 = $2^4 \times 7^2$ [4]

b) Use your answer to part a) to explain why 784 is a square number and then find its square root.

Half of $2^4 \times 7^2$ = 28 witch witch is the square root of 784.

[3]

Total Marks 21 / 21

How do you feel about these skills?

(PS) (MR)

Green = Got it!
Orange = Nearly there
Red = Needs practice

7

Geometry

1. The diagrams below show pairs of equal angles.
   In each case write the word **alternate** or **corresponding** to describe the pair.

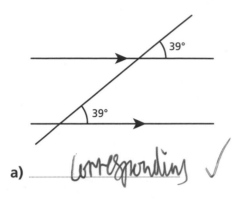

a) _corresponding_ ✓

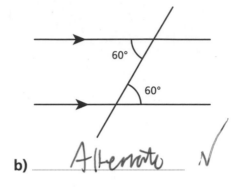

b) _Alternate_ ✓

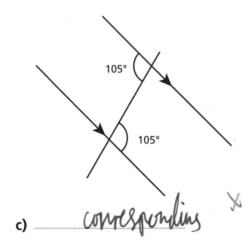

c) _corresponding_ ✗

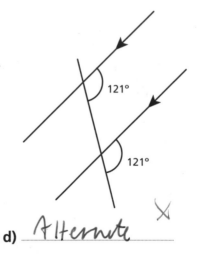

d) _Alternate_ ✗

2 [4]

2. Which **one** of the following statements is always true of a rectangle?

   A: All sides have the same length.

   B: The diagonals intersect at right-angles.

   C: It has rotational symmetry of order 4.

   D: Its diagonals bisect each other.

   E: It has one line of symmetry.

Its diagonals [2] 2
bisect each
other. ✓

3 For the image below, state a possible angle and direction of rotation:

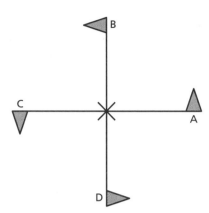

a) from A to B ⟶ 90° | [2]

b) from B to D ⟶ 180° 2 [2]

c) from C to B ⟶ 90° | [2]

d) from D to A ⟶ 90° | [2]

4 Describe the translation:

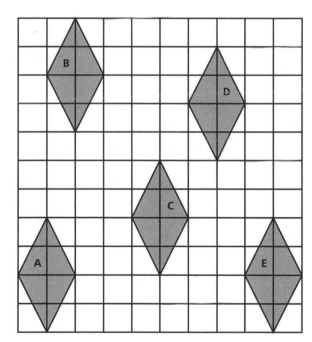

a) from A to B ⟶ 7up , 1 right 2 [2]

b) from B to A ⟶ 7 down , 2 left 2 [2]

c) from E to C ⟶ 4 left , ort 1 up 1 [2]

d) from C to D ⟶ 4 up , 2 right 2 [2]

1 Find the angles marked with letters.

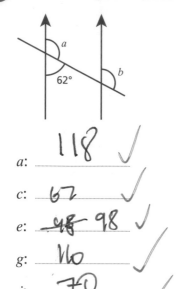 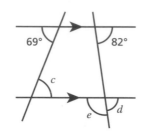 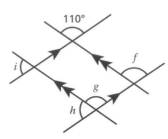

a: __118__ ✓

c: __67__ ✓

e: __~~48~~ 98__ ✓

g: __110__ ✓

i: __70__ ✓

b: __162__ ✗

d: __~~84~~ 84 82__ ✓

f: __110__ ✓
__70__ ✗

_8_[9]

2 Write **True** (T) or **False** (F) in the spaces below.

Property	Parallelogram	Kite
It has one pair of parallel sides	F ✓	F ✓
Its diagonals intersect at right-angles	F ✓	~~T~~ T ✓
It has no lines of symmetry	T ✓	F ✓
It has two pairs of equal angles	T ✓	F ✓

_8_ [8]

3 Use ruler and compasses to construct the perpendicular bisector of the line AB.
Show the construction lines clearly on the diagram.

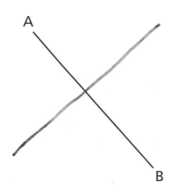

A

B

_1_ [3]

4 Use ruler and compasses to construct the angle bisector of the angle ABC.
Show the construction lines clearly on the diagram.

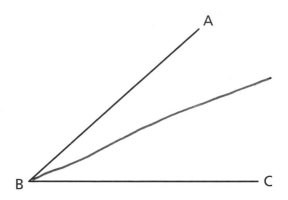

 [3]

5

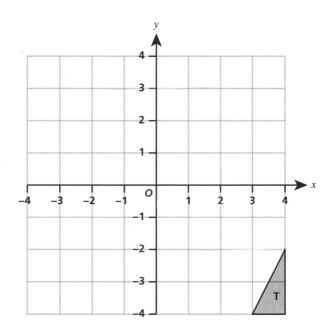

a) Translate triangle T 6 units to the left and 1 unit up.
   Label the image U and write down the coordinates of the vertices of U.

   Coordinates: (-3,-3) (-2,-3) (-2,-1)   [3] 3

b) Translate triangle U 4 units to the right and 5 units up.
   Label the image V and write down the coordinates of the vertices of V.

   Coordinates: (1,2) (2,2)   (2,4)   [3] 3

c) Describe the translation that takes triangle V back to the original triangle T.

   6 down, 2 right   [2] 2

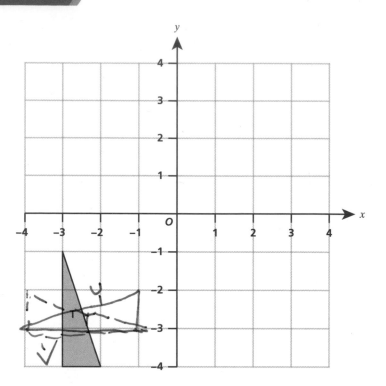

6

a) Rotate triangle T 90° anticlockwise about the origin.

Label the image U and write down the coordinates of the vertices of U.

Coordinates: (1, -3) (4, -3) (4, -2) [3]

b) Rotate triangle U 180° about the origin.

Label the image V and write down the coordinates of the vertices of V.

Coordinates: (-1, 3) (4, 3) (4, 2) [3]

c) Describe a rotation that takes triangle V back to the original triangle T.

Make it not move 90° clockwise [3]

Total Marks 36 / 40

MR **1** An object, A, is moved onto its image, B, by a translation 25 units to the left and 7 units down.

Object B is then moved onto its image, C, by a translation 18 units to the right and 10 units up.

Describe the translation:

a) from A to C [2]

b) from B to A [2]

(MR) **2** The properties of various quadrilaterals are given in the table.

Write the names of all quadrilaterals from the box below that have each property.

| square | rectangle | parallelogram | rhombus | kite | arrowhead | trapezium |

Property	Quadrilateral
Diagonals bisect each other	
Two lines of symmetry	
Rotational symmetry of order two	
One pair of equal angles	
One pair of parallel sides	

[5]

3

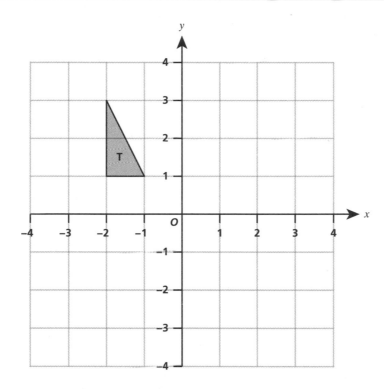

a) Rotate triangle T 90° clockwise about (1, 1).

Label the image U and write down the coordinates of the vertices of U.

Coordinates: .... [3]

b) Translate U 1 unit to the left and 5 units down.

Label the image V and write down the coordinates of the vertices of V.

Coordinates: .... [3]

c) Describe the transformation needed to move V back to the original triangle T.

.... [3]

(MR) **4** Find the angles marked with letters.

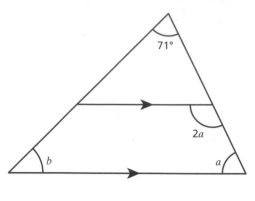

a: _____   b: _____   c: _____   [3]

(PS) **5** The diagram shows a plan view of a school playground.

Ali starts at corner A and begins to walk across the playground. He walks along the straight line that is the angle bisector of AB and AD. He continues along this path until it intersects the perpendicular bisector of CD. He then walks along this perpendicular bisector until he reaches the mid-point of CD.

Draw Ali's route on the diagram showing all construction lines.

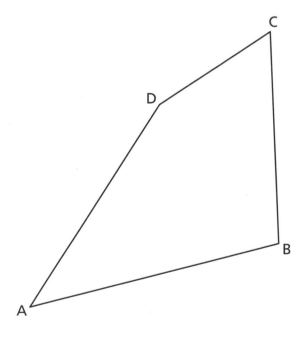

[5]

Total Marks _____ / 26

_____ / 22

_____ / 40

_____ / 26

How do you feel about these skills?

Green = Got it!
Orange = Nearly there
Red = Needs practice

Probability

1 The probability that Kiril scores a goal in the next match is 0.3
What is the probability that he does not score a goal?

.... [1]

2 The probability that a biased coin lands on a 'head' is 0.42
Write down the probability that it lands on a 'tail'.

.... [1]

3 A bag contains 5 counters.

  7

If a counter is picked at random from the bag, what is the probability that it is:

a) an even number .... [1]

b) a number that is a multiple of 3 .... [1]

c) a number that is not a multiple of 3 .... [1]

d) a prime number .... [1]

e) a number that is a factor of 18 .... [1]

f) a number greater than 10 .... [1]

Total Marks .... / 8

1 John experiments by rolling a single dice and a spinner simultaneously.

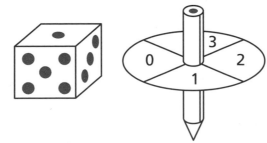

He rolls an ordinary dice with a possible score of 1, 2, 3, 4, 5 or 6. At the same time he also spins a spinner with a possible score of 0, 1, 2 or 3.
His total score is worked out by **multiplying** the two individual scores.

a) Complete the table to show the sample space of final scores. [2]

Score	1	2	3	4	5	6
0						
1			3			
2						12
3				12		

b) Work out the probability of getting a final score of:

   i) 0 _____ [1]

  ii) 7 _____ [1]

 iii) 12 or more _____ [1]

c) Which score is John most likely to get? _____ [1]

2 A drawing pin can either land 'point up' or 'point down'.
Bertie and Ella both throw the same drawing pin in the air and record their results.

Bertie	
Number of throws	40
Point up	15

Ella	
Number of throws	120
Point up	30

a) Write down Bertie's estimate for the probability of the pin landing 'point up'.

_____ [1]

b) Write down Ella's estimate for the probability of the pin landing 'point up'.

_____ [1]

c) Which of these estimates is likely to be the more accurate?
Give a reason for your answer.

_____ [2]

3 A bag contains red and blue counters.
The probability that a red counter is picked at random is 0.4

a) What is the probability that a blue counter is picked? _____ [1]

b) If there are 20 counters altogether in the bag, how many of these are red?

_____ [1]

4 Five events that might occur when a single dice is thrown are:

Event A: get a 6   **Event B:** get less than 3

Event C: get an even number   **Event D:** get a prime number

Event E: get 3 or more

Decide which of the following pairs of events are mutually exclusive.
Write **Yes** if they are and **No** if they are not.

a) A and B ....   b) A and C ....   c) A and D ....

d) A and E ....   e) B and C ....   f) B and D ....

g) B and E ....   h) C and D ....   i) C and E ....

j) D and E ....   [10]

5 Ahmed has three coins in his pocket: 10p, 20p and 50p.
He tosses all three coins at the same time.

a) Complete the table to show the sample space of all eight possible outcomes.   [2]

10p	H	H	H	H	T	T	T	T
20p	H	H	T	T	H	H	......	......
50p	H	T	H	T	......	......	......	......

b) Calculate the probability that when three coins are tossed you get:

i) three heads .... [1]   ii) two heads and a tail .... [1]

6 The letters of the word PROBABILITY are written on 11 cards.
One of these cards is chosen at random.

a) Find the probability of choosing a vowel. ....   [1]

b) Find the probability of choosing the letter 'B'. ....   [1]

c) Find the probability of choosing a card that is either a vowel or the letter 'B'. .... [1]

Total Marks .... / 29

(MR) 1 A bag contains white and black balls.
The probability that a white ball is chosen at random is 0.15

a) What is the probability that a black ball is picked? ....   [1]

b) If there are 9 white balls in the bag, how many black balls are there?

....   [1]

2 A random number generator chooses a whole number between 1 and 20 inclusive.

a) What is the probability that it will choose a factor of 20? _____ [1]

b) What is the probability that it will choose a multiple of 4? _____ [1]

c) What is the probability that it will either choose a factor of 20 or a multiple of 4?

_____ [1]

d) Give a reason why your answer to part c) is **not** the sum of your answers to parts a) and b).

_____ [1]

(MR) 3 Four coins are tossed.

a) In how many possible ways can they land? _____ [2]

b) What is the probability that either they all land heads or they all land tails? _____ [1]

(MR) (PS) 4 A class of 32 students are asked to state whether they like English or Mathematics. Eighteen said they like English, twelve said they like Mathematics and five students said that they didn't like either subject.

a) Complete the Venn diagram to show the preferences of these students. [4]

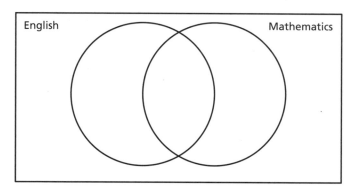

b) If a student is selected at random, what is the probability that they like:

i) both Mathematics and English _____ [1]

ii) at least one of these subjects _____ [1]

iii) English only _____ [1]

Total Marks _____ / 16

_____ / 8

_____ / 29

_____ / 16

How do you feel about these skills?

Green = Got it!
Orange = Nearly there
Red = Needs practice

Percentages

1. Without using a calculator write each of these test marks as a percentage. 📱

   a) 7 out of 14 .... [1]

   b) 30 out of 40 .... [1]

   c) 8 out of 10 .... [1]

   d) 31 out of 50 .... [1]

2. Write down the multiplier associated with each percentage increase.

   a) 20% ....   b) 5% ....   c) 50% .... [3]

3. Find the percentage increase associated with each of these multipliers.

   a) 1.03 ....   b) 1.32 ....   c) 1.015 .... [3]

   Total Marks .... / 10

---

1. Write down the multiplier associated with each percentage decrease.

   a) 25% ....   b) 10% ....   c) 2% .... [3]

2. Find the percentage decrease associated with each of these multipliers.

   a) 0.85 ....   b) 0.4 ....   c) 0.995 .... [3]

(PS) 3. Use the multiplier method to find the new quantities when:

   a) an annual salary of £25 000 rises by 4 percent

   .... [1]

   b) the population of a village of 2600 rises by 25 percent

   .... [1]

   c) an elastic band 6cm long is stretched by 15 percent

   .... [1]

   d) the price of a dress costing £160 is reduced by 20 percent in a sale

   .... [1]

e) a person with a mass of 110kg loses 8 percent on a diet

.... [1]

f) the number of rabbits, 45 000, is reduced by 12 percent

.... [1]

(FS) **4** The average price of a house in the UK at the start of 2013 was £161 500.
At the start of 2014 this had risen to £175 000.

Work out the percentage increase in house prices during 2013.
Give your answer to one decimal place.

....

.... [3]

(FS) **5** A coat originally priced at £240 is reduced to £156 in a sale.
Work out the percentage decrease in price.

....

.... [3]

Total Marks .... / 18

1 Write down the percentage change associated with each of these multipliers.
In each case, state whether it is an **increase** or a **decrease**.

a) 1.175 .... [1]

b) 2.5 .... [1]

c) 0 .... [1]

(FS) (MR) (PS) **2** Train fares rise in January each year.
The cost of an annual season ticket from Surrey to London is shown in the table.

Year	January prices
2011	£3300
2012	£3550
2013	£3763
2014	£3875

a) Work out the annual percentage increases in rail fares in each year.
Round your answers to one decimal place.

    i) 2012 .... [1]

    ii) 2013 .... [1]

    iii) 2014 .... [1]

b) Work out the overall percentage increase from 2011 to 2014.

    .... [1]

c) Explain why the answer to part **b)** is not just the sum of the three answers in part **a)**.

    ....

    .... [1]

3 **a)** A hotel room costs £200 a night, excluding a 12 percent tax.

    i) Work out the price of the room when the tax is included.

    .... [2]

    ii) In an attempt to offset the tax the hotel now reduces its prices by 12 percent.
Work out the new price of the room and explain why this reduction has failed to
restore the price to its original value of £200.

    ....

    .... [3]

b) A new spending tax of 25 percent is imposed on all electrical goods. A shop decides to
reduce all of its electrical goods in a sale.
What percentage reduction is needed to restore all prices to their pre-tax values?

> Consider what happens to the price of an item that costs £100 before the tax is
> imposed.

    ....

    ....

    .... [3]

Total Marks .... / 16

.... / 10

.... / 18

.... / 16

Sequences

1. Write down numbers generated by each of the following flow charts.

   a)

   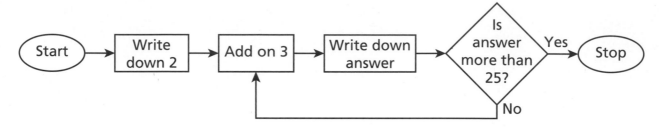

   .... [3]

   b)

   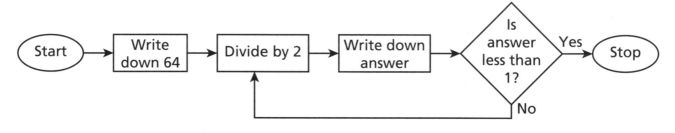

   .... [3]

2. Find the first, second, third and tenth terms of sequences whose nth term is given by:

   a) $4n + 1$ .... [4]

   b) $50 - 2n$ .... [4]

3. Find the first three terms of a sequence that has a first term of seven and a difference of five.

   .... [3]

4. Find the next two terms in each of these Fibonacci sequences.

   a) 4, 5, 9, 14, ... , ... [2]

   b) 1, 6, 7, 13, ... , ... [2]

Total Marks .... / 21

1 Work out the nth term for each of the following sequences.

a) 5, 10, 15, 20, 25, ... [2]

b) 4, 6, 8, 10, 12, ... [2]

c) 12, 15, 18, 21, 24, ... [2]

d) 17, 15, 13, 11, 9, ... [2]

e) 100, 90, 80, 70, 60, ... [2]

f) 4, 4.5, 5, 5.5, 5 .... [2]

g) 8.9, 8.8, 8.7, 8.6, 8.5, ... [2]

(PS) **2** Look at the following patterns made from matchsticks.

Pattern 1   **Pattern 2**   **Pattern 3**

a) Write down the number of matchsticks in the nth pattern.

.... [2]

b) How many matchsticks are there in the 100th pattern?

.... [1]

(PS) **3** A sunflower is 15cm tall on 1st July.

During the month it grows 2cm a day so that on 2nd July it is 17cm tall and on 3rd July it is 19cm tall.

a) Write down an expression for the height on the nth day of July.

.... [2]

b) How tall is the sunflower on 31st July?

.... [1]

c) On which date is the sunflower 33cm tall?

.... [1]

(PS) **4** Here is a pattern of grey and white tiles.

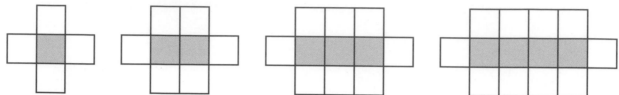

a) Complete the table to show the numbers of grey and white tiles. [1]

Grey	1	2	3	4
White				

b) If a shape has n grey tiles write down a formula for the number of white tiles.

.... [2]

c) Explain why there is no shape in this pattern that has 79 white tiles.

.... [1]

d) Write down a formula for the **total** number of grey and white tiles for a shape that has n grey tiles.

.... [2]

Total Marks .... / 27

(MR) **1** Fill in the missing numbers in these Fibonacci sequences.

a) –2, –1, ..., ... [2]

b) 13, ..., 47, ... [2]

c) ...., 31, ..., ..., 109 [3]

(MR) **2** Write down the nth term of each of the following sequences.

a) 1, 4, 9, 16, 25, ... [1]

b) 0, 3, 8, 15, 24, ... [1]

c) 0, 1, 4, 9, 16, ... [1]

d) 1, 8, 27, 64, 125, ... [1]

e) 3, 10, 29, 66, 127, ... [1]

3 Here is a sequence of squares.

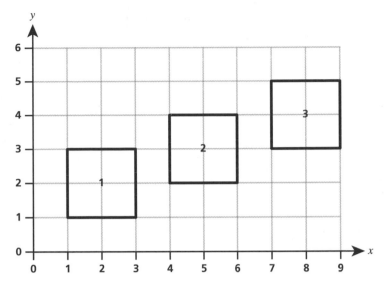

a) Find an expression for each of the x and y coordinates of the centre of the nth square.

....

....

(.............................. , ...) [4]

b) Hence, or otherwise, write down the:

i) x-coordinate of the centre of the square that has a y-coordinate of 41

....

$x =$ .... [2]

ii) coordinates of the top left-hand corner of the nth square

....

....

(.............................. , ...) [4]

Total Marks .... / 22

.... / 21

.... / 27

.... / 22

How do you feel about these skills?

Green = Got it!
Orange = Nearly there
Red = Needs practice

Area of 2D and 3D Shapes

1 Write down the name of each of these shapes.

a)

b)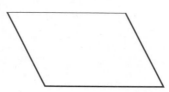

c)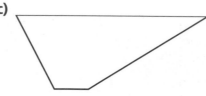

....   [3]

PS **2** Work out the areas of these triangles.

a)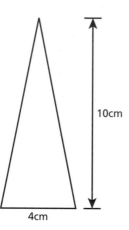

10cm

4cm

b)

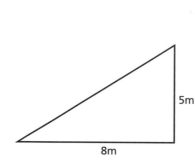

5m

8m

c)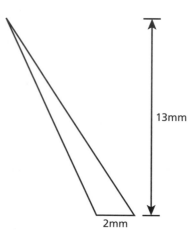

13mm

2mm

....
.... cm²   .... m²   .... mm²   [6]

PS **3** Work out the areas of these parallelograms.

a)

3cm

7cm

b)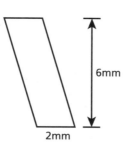

6mm

2mm

c)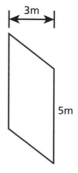

3m

5m

....
.... cm²   .... mm²   .... m²   [6]

4 Work out the areas of these trapezia.

a)

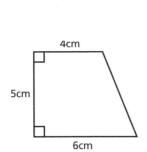

b)

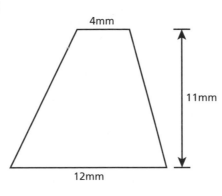

c)

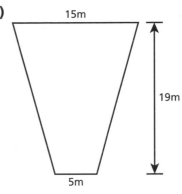

....

....

.... cm²   .... mm²   .... m²   [6]

1 Work out the surface areas of these cuboids.

a)

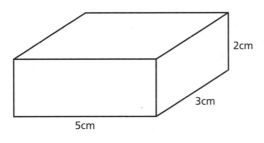

b)

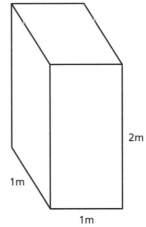

....

....

.... cm²   .... m²   [6]

(PS) **2** Find the area of each of the following shapes.

a)

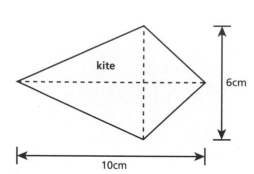

b)

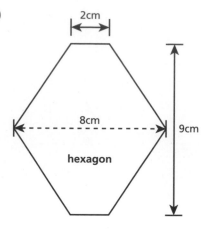

....

....

....

.... cm²

....

....

....

.... cm²   [6]

(PS) **3** The diagram shows an area of woodland.

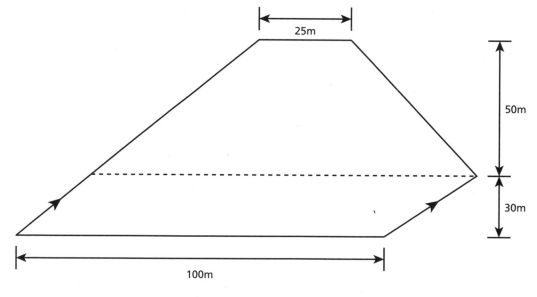

a) Calculate the total area of woodland.

....

....

.... m² [5]

(FS) b) Work out the value of this land if it sells for £190 per 100m².

.... [2]

(4) Find x.

a)

10cm

Area = 60cm²

....

....

.... cm

b)

x

4m

Area = 16m²

....

....

.... m

c)

1mm

x

5mm

Area = 18mm²

....

....

.... mm   [6]

Total Marks .... / 25

(PS)
(MR) (1) The surface area of a cube is 150cm².

Find the length of each edge.

....

....

.... cm [3]

(PS)
(MR) (2) The surface area of the cuboid shown below is 128cm².

Find the length x.

Give full details of your working.

$x$cm

4cm

4cm

....

....

....

....

....   [4]

(PS)
(MR)

3 The diagram shows a shaded square drawn inside a larger square with sides of length 10cm.
Find the area of the shaded square.

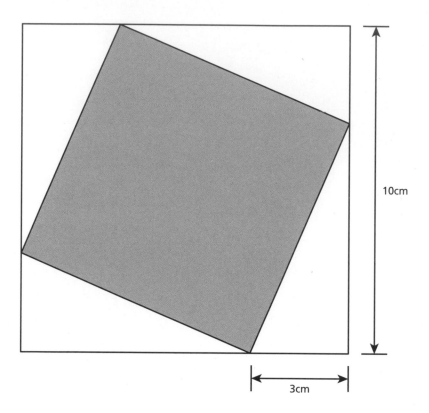

10cm

3cm

Shaded area = ... cm² [4]

.... / 21

.... / 25

.... / 11

How do you feel about these skills?

Green = Got it!
Orange = Nearly there
Red = Needs practice

Graphs

1 State the gradient of each of these lines:

    a) $y = 5x + 3$ **b)** $y = 2x + 9$ **c)** $y = x + 4$

    ....   **[3]**

2 **a)** Fill in the missing numbers in the table for $y = 2x + 3$ **[2]**

x	−2	−1	0	1	2	3
$y = 2x$	−4	−2	0		4	
$y = 2x + 3$	−1	1	3		7	

    b) Plot these points on the grid and draw the graph of $y = 2x + 3$.   **[2]**

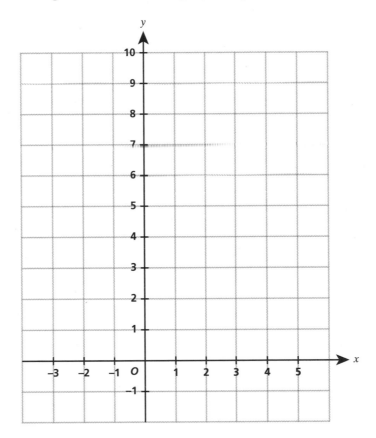

3 State the equation of the straight line with:

    a) a gradient of 3 passing through (0, 2) ....   **[2]**

    b) a gradient of 7 passing through (0, 1) ....   **[2]**

    c) a gradient of 2 passing through (0, 7) ....   **[2]**

Total Marks .... **/ 13**

1 For each of these graphs:

a) find the gradient of the line

b) write down the coordinates of where the line crosses the y-axis

c) write down the equation of the line

A

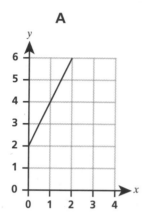

a) ....

b) ....

c) ....

B

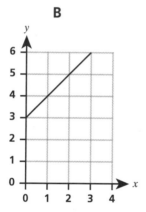

a) ....

b) ....

c) ....

C

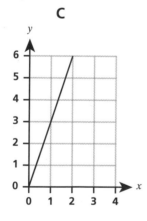

a) ....

b) ....

c) ....

[9]

2 **a)** Fill in the missing numbers in the table for $y = x^2 + 6$. [5]

x	−3	−2	−1	0	1	2	3
x^2		4			1		
$y = x^2 + 6$	....	10	....	....	7	....	....

b) Plot these points on the grid and draw the graph of $y = x^2 + 6$. **[4]**

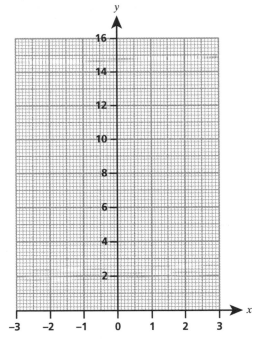

3 Jenny cycles from her house to the post office and back again.
The distance-time graph of her journey is shown below.

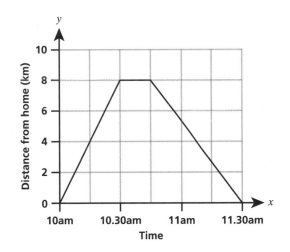

a) At what time does Jenny leave her house? .... **[1]**

b) At what speed does she cycle to the post office?

.... km/h **[1]**

c) How many minutes does she spend at the post office?

.... min **[1]**

d) What is the speed of her return journey? Give your answer as a mixed number.

.... km/h **[1]**

Total Marks .... / 22

(MR) 1 Write down the equation of the straight line that is parallel to the line $y = 4x + 7$ and passes through the point (0, 5).

.... [2]

(PS) 2 A ball is dropped from the top of a tall building onto the ground below.

The distance, d, metres travelled by the ball t seconds after it is dropped is given by $d = 4t^2$.

The ball hits the ground after four seconds.

a) Fill in the table to show the distance fallen. [5]

t	0	1	2	3	4
t^2					
$d = 4t^2$					

b) Plot these points on the grid and draw a graph of $d = 4t^2$. [4]

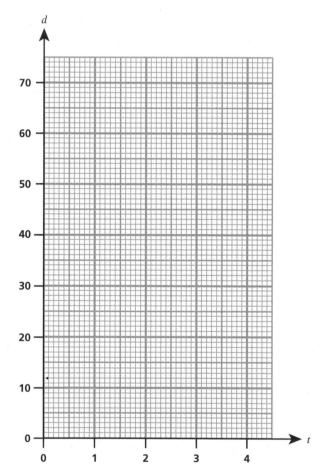

c) How tall is the building? .... m [1]

d) How long does it take for the ball to fall 25m from the top of the building?

.... s [2]

3 A car travels the 200 miles along the M1 from London to Leeds. The car leaves London at 8am and travels at a constant speed of 60mph for the first 60 miles.

The driver takes a break of 30mins at a motorway services before continuing the journey at a constant speed of 70mph.

a) Draw a distance-time graph of the journey. [3]

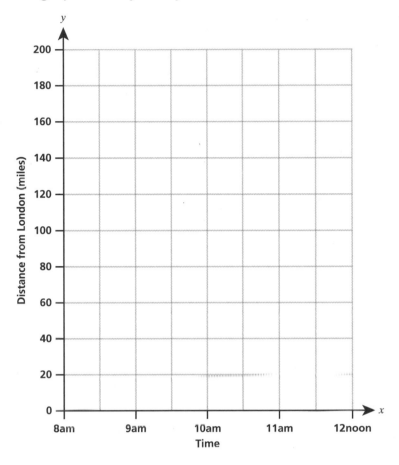

b) Work out the average speed for the complete journey.
Give your answer as a mixed number.

_____ [2]

c) A lorry leaves Leeds at 8am and travels to London without stopping at a constant speed of 50mph. Draw the distance-time graph of the lorry on the same grid. [1]

d) At what time do the car and lorry pass each other? _____ [1]

Total Marks _____ / 21

	/ 13
	/ 22
	/ 21

How do you feel about these skills?

Simplifying Numbers

1 Work out each of the following.

a) $2.52 \times 10 =$ ....

b) $0.000452 \times 100 =$ ....

c) $9.4 \div 10 =$ ....

d) $5.42 \div 1000 =$ ....

e) $0.000769 \times 10^4 =$ ....

f) $0.000654 \times 10^8 =$ ....

g) $234100 \div 10^4 =$ ....

h) $734 \div 10^5 =$ ....   [8]

2 Cross out the numbers in this list that are **not** in standard form.

4×10^3 0.6×10^9 6.52×10^2 12×10^7

9.9×10^{17} 0.99×10^6 1.1×10^{99} [3]

Total Marks .... / 11

1 Round each of the following to the number of significant figures indicated.

a) 5.73 (2 sf)

b) 451 763 (3 sf)

c) 17.03 (1 sf)

d) 300 965 (4 sf)

....

e) 0.736 (1 sf)

f) 0.08067 (3 sf)

g) 0.0000955 (2 sf)

h) 0.8069 (2 sf)

....   [8]

2 Round 6 872 430 to the nearest a) million b) hundred thousand c) ten thousand.

a) ....   b) ....   c) ....   [3]

3 Express each of the following numbers in standard form.

a) 74

b) 965 000

c) 2.5 million

d) 1358

....   [4]

4 Without using a calculator, work out each of the following. 🚫🖩
Give your answers in standard form.

a) $(2 \times 10^3) \times (4 \times 10^2) =$ ....

b) $(3 \times 10^3) \times (2 \times 10^5) =$ ....

c) $(3.5 \times 10^6) \times (2 \times 10^5) =$ ....

d) $(3 \times 10^4)^2 =$ ....   [4]

Answers

Pages 4–7

Working with Numbers

1. a) −20 [1] b) 21 [1] c) −8 [1] d) −2 [1] e) −2 [1] f) 4 [1]
   g) −35 [1] h) 2 [1] i) 0 [1] j) 25 [1] k) −42 [1] l) −4 [1]
2. a) 1, 2, 4, 8 [2 marks: −1 per extra/omission]
   b) 1, 2, 3, 4, 6, 12 [2 marks: −1 per extra/omission]
   c) 1, 2, 4 [1 mark for all three correct values]
   d) 4   [1]
3. a) 6, 12, 18, 24, 30 [2 marks: −1 per extra/omission]
   b) 10, 20, 30, 40, 50 [2 marks: −1 per extra/omission]
   c) 30   [1]
4. Circled numbers: 2, 3, 5, 7, 11 [2 marks: −1 per extra/omission]
5. a) 128 [1] b) 243 [1] c) 256 [1]
6. a) 21 [1] b) 6 [1] c) 8.5 [1]

1. a) 3 [1] b) −6 [1] c) −12 [1] d) 6 [1]
2. 2, 3, 7 [2 marks: −1 per extra/omission]
3. 53   [2]
4. 9:10am [3 marks: 1 for attempt to find LCM; 1 for the value
   of 60; 1 for correct time]
5. a) $3 \times 3 \times 7$ [3 marks: 1 for each correct prime factor]
   b) $2 \times 5 \times 7$ [3 marks: 1 for each correct prime factor]
6. [2 marks for each part below: −1 for any extras/omissions]
   a) 1, 2, 3, 4, 6, 8, 12, 16, 24, 48
   b) 1, 2, 3, 4, 6, 9, 12, 18, 36
   c) 1, 2, 3, 6, 9, 18
   d) 1, 2, 4, 8, 16, 32, 64
   e) 1, 3, 5, 9, 15, 45

   > To avoid missing out some of the factors it helps if you write
   > them down in pairs, e.g. In a) 1, 48; 2, 24; 3, 16; 4, 12; 6, 8.

7. a) 12 [1] b) 18 [1] c) 16 [1] d) 9 [1]
8. a) ±13 [2] b) ±1.2 [2] c) ±47 [2] [2 marks: 1 for obtaining
   two answers, one positive and the other negative; 1 for the
   numerical value]

1. a) −24 [1] b) 4 [1] c) 2 [1]
2. a) HCF = 40 [1] so 12:40am [1]
   b) HCF = 120 [1] so 2am [1]
   c) 3   [1]
3. a) −8 and −3 [1] b) 16 and −3 [1]
4. a) $2^5 \times 3^2 \times 5^2 \times 7^4$ [2] b) $2^6 \times 3^4 \times 5^3 \times 7^4$ [2] [Marks only
   awarded for completely correct answer]
5. a) $2^4 \times 7^2$ [4 marks: 2 for using either the factor tree or division
   methods correctly; 1 for correct power of 2; 1 for correct
   power of 7]
   b) Accept suitable answer, e.g. Powers are both even. [3 marks:
   1 for 'Powers are both even; 1 for halving the powers of
   a) to get $2^2 \times 7$; 1 for correct answer of 28. No marks given
   unless the result of part a) has been used]

Pages 8–14

Geometry

1. a) corresponding [1] b) alternate [1] c) alternate [1]
   d) corresponding [1]
2. D   [2]
3. a) 90° anti-clockwise [2] b) 180° [2] c) 90° clockwise [2]
   d) 90° anti-clockwise [2]
4. a) 1 right, 7 up b) 1 left, 7 down c) 4 left, 2 up

d) 2 right, 4 up [2 marks in each part: 1 for left/right instruction
   and 1 for up/down]

1. a: = 118° [1] b: = 118° [1] c: = 69° [1] d: = 82° [1] e: = 98 [1]
   f: = 110° [1] g: = 110° [1] h: = 70° [1] i: = 70° [1]

2.

Property	Parallelogram	Kite
It has one pair of parallel sides	F [1]	F [1]
Its diagonals intersect at right-angles	F [1]	T [1]
It has no lines of symmetry	T [1]	F [1]
It has two pairs of equal angles	T [1]	F [1]

3.

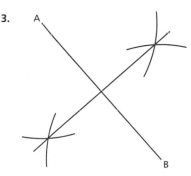

[3 marks: 1 for each pair of arcs; 1 for line]

4.

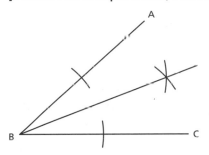

[3 marks: 2 for construction lines; 1 for final answer]

5. a) (−3, −3) [1] (−2, −3) [1] (−2, −1) [1]
   b) (1, 2) [1] (2, 2) [1] (2, 4) [1]
   c) 2 right [1] 6 down [1]
6. a) (1, −3) [1] (4, −3) [1] (4, −2) [1]
   b) (−1, 3) [1] (−4, 3) [1] (−4, 2) [1]
   c) 90° [1] anticlockwise [1] about (0, 0) [1]

1. a) 7 left [1] 3 up [1] b) 25 right [1] 7 up [1]

2.

Property	Quadrilateral
Diagonals bisect each other	Square, rectangle, parallelogram, rhombus
Two lines of symmetry	Rectangle, rhombus
Rotational symmetry of order two	Rectangle, rhombus, parallelogram
One pair of equal angles	Kite, arrowhead
One pair of parallel sides	Trapezium

[5 marks: 1 for each correct list]

3. a) (1, 3) [1] (1, 4) [1] (3, 4) [1]

b) (0, −2) [1] (0, −1) [1] (2, −1) [1]

c) rotation [1] 90° anticlockwise [1] about (−2, −1) [1]

4. a: = 60° [1] b: = 49° [1] c: = 43° [1]

5.

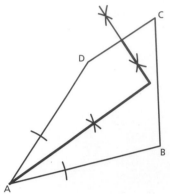

[5 marks: 1 for the angle bisector; 1 for its construction lines; 1 for the perpendicular bisector; 1 for its construction lines; 1 for the complete route.]

Pages 15–18

Probability

1. 0.7 [1]
2. 0.58 [1]
3. a) $\frac{2}{5}$ [1] b) $\frac{3}{5}$ [1] c) $\frac{2}{5}$ [1] d) $\frac{2}{5}$ [1] e) $\frac{3}{5}$ [1] f) 0 [1]

1. a)

Score	1	2	3	4	5	6
0	0	0	0	0	0	0
1	1	2	3	4	5	6
2	2	4	6	8	10	12
3	3	6	9	12	15	18

[2 marks for correct grid; −1 for any error]

b) i) $\frac{1}{4}$ [1] ii) 0 [1] iii) $\frac{1}{6}$ [1]

c) 0 [1]

2. a) $\frac{3}{8}$ [1]

b) $\frac{1}{4}$ [1]

c) Ella [1] since she has more throws. [1]

3. a) 0.6 [1]

b) 8 [1]

4. a) Y [1] b) N [1] c) Y [1] d) N [1] e) N [1] f) N [1] g) Y [1]

h) N [1] i) N [1] j) N [1]

5. a)

10p	H	H	H	H	T	T	T	T
20p	H	H	T	T	H	H	T	T
50p	H	T	H	T	H	T	H	T

[2 marks for correct grid; −1 for any error]

b) i) $\frac{1}{8}$ [1] ii) $\frac{3}{8}$ [1]

6. a) $\frac{4}{11}$ [1] b) $\frac{2}{11}$ [1] c) $\frac{6}{11}$ [1]

If the word 'or' occurs in a probability question, make sure that the events do not overlap before you add probabilities. It is only when the events are mutually exclusive that you can do so.

1. a) 0.85 [1]

b) 51 [1]

In part b) you need to solve the equation 0.15n = 9 to first work out the total number of balls, n.

2. a) $\frac{3}{10}$ [1] b) $\frac{1}{4}$ [1] c) $\frac{9}{20}$ [1]

d) not mutually exclusive (4 and 20 are common) [1]

3. a) 16 [2] b) $\frac{1}{8}$ [1]

You can either write out the table as in Q5 of Probability 2 or notice that there are two outcomes for each of the four coins so the number of ways four coins can land is 2 × 2 × 2 × 2.

4. a)

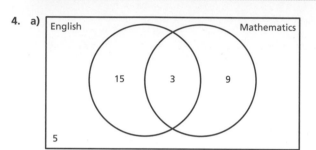

[4 marks: 1 for each figure]

b) i) $\frac{3}{32}$ [1] ii) $\frac{27}{32}$ [1] iii) $\frac{15}{32}$ [1]

Pages 19–21

Percentages

1. a) 50% [1] b) 75% [1] c) 80% [1] d) 62% [1]
2. a) 1.2 [1] b) 1.05 [1] c) 1.5 [1]
3. a) 3% [1] b) 32% [1] c) 1.5% [1]

1. a) 0.75 [1] b) 0.9 [1] c) 0.98 [1]
2. a) 15% [1] b) 60% [1] c) 0.5% [1]
3. a) £26 000 [1] b) 3250 [1] c) 6.9cm [1] d) £128 [1]

e) 101.2kg [1] f) 39 600 [1]

4. 8.4% [3 marks for correct answer with no working; however if final answer is wrong then 1 for dividing 175 000 by 161 500; 1 for 1.083591]

5. 35% [3 marks for correct answer with no working; however if final answer is wrong then 1 for dividing 156 by 240; 1 for 0.65]

1. a) 17.5% increase [1]

b) 150% increase [1]

c) 100% decrease [1]

2. a) i) 7.6% [1] ii) 6.0% [1] iii) 3.0% [1]

b) 17.4% [1]

c) Accept suitable answer, e.g. 'The 6 percent increase in 2013 not only applies to the original fare of £3300 in 2011 but also to the increase of 7.6 percent (in fact £250) from the previous year'. [1]

3. a) i) £224 [2]

ii) £197.12 [1] Accept suitable answer, e.g. 'The increase is 12 percent of £200 which is not the same as the decrease which is 12 percent of £224'. [2]

b) 20% [3 marks for correct answer with no working; however if final answer is wrong then an answer such as the following argument scores individual marks]. 'Assume that the electrical item is £100 before the tax. After the tax this rises to £125. [1] To restore this to £100 the multiplier would be 100/125 = 0.8 [1] which represents a 20 percent reduction' [1].

Sequences

1. a) 2, 5, 8, 11, 14, 17, 20, 23, 26 [3 marks: –1 per extra/incorrect number]
   b) 64, 32, 16, 8, 4, 2, 1, 0.5 [3 marks: –1 per extra/incorrect number]
2. a) 5 [1] 9 [1] 13 [1] 41 [1]
   b) 48 [1] 46 [1] 44 [1] 30 [1]
3. 7 [1] 12 [1] 17 [1]
4. a) 23 [1] 37 [1]
   b) 20 [1] 33 [1]

1. [2 marks for each correct answer: 1 for the coefficient of n and 1 for number term]
   a) $5n$ [2] b) $2n + 2$ [2] c) $3n + 9$ [2] d) $19 - 2n$ [2]
   e) $110 - 10n$ [2] f) $0.5n + 3.5$ [2] g) $9 - 0.1n$ [2]
2. a) $2n + 1$ [2 marks: 1 for $2n$; 1 for +1]
   b) 201 [1]
3. a) $2n + 13$ [2 marks: 1 for $2n$; 1 for +13]
   b) 75cm [1]
   c) 10th July [1]

4. a)

Grey	1	2	3	4
White	4	6	8	10

   [1 mark for correct table]
   b) $2n + 2$ [2 marks: 1 for $2n$; 1 for +2]
   c) Acceptable suitable answer e.g. '79 is not even' [1]
   d) $3n + 2$ [2 marks: 1 for $3n$; 1 for +2]

Either add a third row to the table for the total or simply add n to the answer to part b).

1. a) –3 [1] –4 [1]
   b) 34 [1] 81 [1]
   c) 8 [1] 39 [1] 70 [1]

In parts b) and c) it helps if you let x be the first unknown term in the sequence and follow it through the calculations e.g. in c): x, 31, $x + 31$, $x + 62$, $2x + 93$ so $2x + 93 = 109$.

2. a) n^2 [1] b) $n^2 - 1$ [1] c) $(n - 1)^2$ [1] d) n^3 [1] e) $n^3 + 2$ [1]

In these sequences the numbers are getting further and further apart. When this happens write down the first few square and cube numbers and then try adjusting them to fit the given sequence.

3. a) $(3n - 1, n + 1)$ [4 marks: 1 for each of $3n$, –1, n, +1]
   b) i) 119 [2 marks for correct answer; otherwise give 1 for sight of $n = 40$]
      ii) $(3n - 2, n + 2)$ [4 marks: 1 for each of $3n$, –2, n, +2]

Subtract 1 from the x-coordinate and add 1 to the y-coordinate of the answer to the first part of the question.

Area of 2D and 3D Shapes

1. a) trapezium [1] b) parallelogram [1] c) trapezium [1]
2. a) $\frac{1}{2} \times 4 \times 10$ [1] = 20cm² [1] b) $\frac{1}{2} \times 8 \times 5$ [1] = 20m² [1]
   c) $\frac{1}{2} \times 2 \times 13$ [1] = 13mm² [1]
3. a) 7×3 [1] = 21cm² [1] b) 2×6 [1] = 12mm² [1]
   c) 3×5 [1] = 15m² [1]

4. a) $\frac{1}{2}(4 + 6) \times 5$ [1] = 25cm² [1]
   b) $\frac{1}{2}(4 + 12) \times 11$ [1] = 88mm² [1]
   c) $\frac{1}{2}(15 + 5) \times 19$ [1] = 190m² [1]

1. a) $2 \times (2 \times 5) = 20$, $2 \times (2 \times 3) = 12$, $2 \times (3 \times 5) = 30$
      20 + 12 + 30 = 62cm² [3]
   b) $2 \times (2 \times 1) = 4$, $2 \times (1 \times 2) = 4$, $2 \times (1 \times 1) = 2$
      4 + 4 + 2 = 10m² [3]
   [3 marks: 2 accuracy; 1 method]
2. a) $\frac{1}{2} \times 10 \times 3 = 15$, $15 \times 2 = 30$cm² [3]
   b) $\frac{1}{2}(2 + 8) \times 4.5 = 22.5$, $22.5 \times 2 = 45$cm² [3]
   [3 marks: 2 accuracy; 1 method]

Divide compound shapes into simpler shapes. The kite consists of two identical triangles one on top of the other. The hexagon consists of two trapezia.

3. a) area of parallelogram is 3000m² [2]; area of trapezium is 3125m² [2]; giving total area 6125m² [1].
   b) £11 637.50 [2 marks: 1 for attempting to multiply answer to a) by £190; 1 for answer]
4. a) 6cm [2] b) 8m [2] c) 6mm [2]
   [2 marks for correct answers: if any of these values are wrong then 1 for attempting to find areas e.g. for writing down the area of the parallelogram as $10x$ in a), the area of the triangle as $2x$ in b) and the area of the trapezium as $3x$ in c)]

1. $6x^2 = 150$ or $x^2 = 25$ so $x = 5$ [3 marks: 2 for either $6x^2 = 150$ or $x^2 = 25$; 1 for final answer of 5]
2. $2(16 + 8x) = 128$ or $16 + 8x = 64$ [2] so $x = 6$ [1] [4 marks: 2 for either $2(16 + 8x) = 128$ or $16 + 8x = 64$; 1 for attempting to solve either equation; 1 for final answer of 6]
3. 58; large square has area 100 and each of the four right-angled triangles has area $\frac{1}{2} \times 3 \times 7$ so the shaded area is
   $100 - 4\left(\frac{1}{2} \times 3 \times 7\right) = 58$ [4 marks: 1 for the area of 100, 1 for the area of each triangle; 1 for overall method; 1 for final answer of 58]

The area of a shape that has been rotated like this can often be found by surrounding it by a shape whose area is easy to find and then cutting off the triangular corners.

Graphs

1. a) 5 [1] b) 2 [1] c) 1 [1]

2. a)

x	–2	–1	0	1	2	3
$y = 2x$	–4	–2	0	2	4	6
$y = 2x + 3$	–1	1	3	5	7	9

   [2 marks: 1 for 5; 1 for 9]

Check that you agree with the numbers already given in the table. This tells you that your own calculations are likely to be correct.

b)

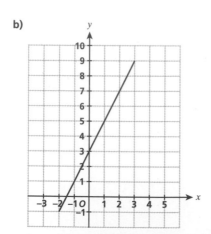

[2 marks: 1 for plotting points correctly; 1 for line]
3. a) $y = 3x + 2$ [2 marks: 1 for x term; 1 for constant term]
    b) $y = 7x + 1$ [2 marks: 1 for x term; 1 for constant term]
    c) $y = 2x + 7$ [2 marks: 1 for x term; 1 for constant term]

Remember to write $y = ...$ to avoid losing marks.

1. A: a) 2 [1] b) (0,2) [1] c) $y = 2x + 2$ [1]
    B: a) 1 [1] b) (0,3) [1] c) $y = x + 3$ [1]
    C: a) 3 [1] b) (0,0) [1] c) $y = 3x$ [1]

2. a)

x	–3	–2	–1	0	1	2	3
x^2	9	4	1	0	1	4	9
$y = x^2 + 6$	15	10	7	6	7	10	15

[5 marks: 1 for each of the five numbers on the bottom row]

As a check on your calculations notice that the numbers are symmetrically placed in the table.

b) [4 marks: 3 for plotting points accurately; –1 per error; 1 for joining the points with a smooth curve]

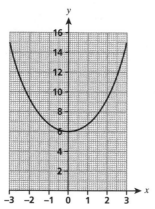

3. a) 10am [1] b) 16km/h [1] c) 15mins [1] d) $10\frac{2}{3}$km/h [1]

1. $y = 4x + 5$ [2 marks: 1 for each term]

2. a)

t	0	1	2	3	4
t^2	0	1	4	9	16
$d = 4t^2$	0	4	16	36	64

[5 marks: 1 for each of the five numbers in the bottom row]

b)

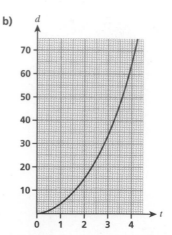

[4 marks: 3 for plotting points accurately; –1 per error; 1 for joining the points with a smooth curve]
c) 64m   [1]
d) 2.5s [2 marks: 1 for an answer to within 0.5s; 2 to within 0.25s]

3. a) [3 marks: 1 for each of the three sections of the car journey]

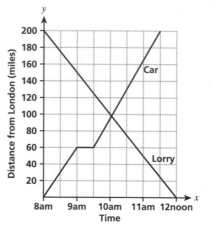

b) $\frac{200}{3.5} = 57\frac{1}{7}$ [2 marks: 1 for (total distance/time); 1 for numerical value]
c) Lorry journey line. See graph above.   [1]
d) Accept answer between 10am and 10:05am   [1]

Pages 36–37

Simplifying Numbers

1. a) 25.2 [1] b) 0.0452 [1] c) 0.94 [1] d) 0.00542 [1] e) 7.69 [1]
    f) 65400 [1] g) 23.41 [1] h) 0.00734 [1]
2. Cross out numbers: 0.6×10^9, 12×10^7, 0.99×10^6 [3 marks; –1 per omission/extra]

You just need to check that the number part is between 1 and 10 so can ignore the power.

1. a) 5.7 [1] b) 452000 [1] c) 20 [1] d) 301000 [1] e) 0.7 [1]
    f) 0.0807 [1] g) 0.000096 [1] h) 0.81 [1]
2. a) 7000000 [1] b) 6900000 [1] c) 6870000 [1]
3. a) 7.4×10^1 [1] b) 9.65×10^5 [1] c) 2.5×10^6 [1]
    d) 1.358×10^3 [1]
4. a) 8×10^5 [1] b) 6×10^8 [1] c) 7×10^{11} [1] d) 9×10^8 [1]
5. a) 7.6×10^{10} [1] b) 5.7×10^6 [1] c) 1.5×10^{19} [1]

1. a) 2.4×10^9 [1] b) 1.2×10^{18} [1] c) 3×10^{14} [1]
    d) 4.9×10^{21} [1] e) 6.4×10^{19} [1]

2. a) $1.429\,624 \times 10^{12}$ [1]
   b) £54 000 [2 marks: 1 for dividing by 26.5 million; 1 for rounding]
3. No. of seconds in a year = $365 \times 24 \times 60 \times 60 = 3.1536 \times 10^7$ [1]
   Light travels = $3.1536 \times 10^7 \times 3 \times 10^8 = 9.4608 \times 10^{15}$m in a year [1]
   56 light years = $56 \times 9.4608 \times 10^{15} = 5.3 \times 10^{17}$m [1]
   = 5.3×10^{14}km [1]

Pages 38–42

Interpreting Data

1. a) No correlation [1]   b) Negative [1]   c) Positive [1]
2. a) 6 [1]
   b) 8 [1]
3. Example:

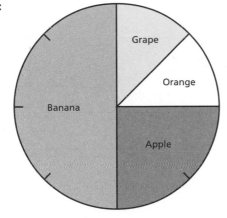

[3 marks: 1 for the correct size of each sector]

1.

Test Result	Angle	Fraction	Number of boys
D	30°	$\frac{30}{360} = \frac{1}{12}$ [1]	$\frac{1}{12} \times 72 = 6$ [1]
C	60°	$\frac{60}{360} = \frac{1}{6}$	$\frac{1}{6} \times 72 = 12$
B	50°	$\frac{50}{360} = \frac{5}{36}$ [1]	$\frac{5}{36} \times 72 = 10$ [1]
A	120°	$\frac{120}{360} = \frac{1}{3}$ [1]	$\frac{1}{3} \times 72 = 24$ [1]
A*	100°	$\frac{100}{360} = \frac{5}{18}$ [1]	$\frac{5}{18} \times 72 = 20$ [1]

2. a) positive [1]   b) positive [1]   c) negative [1]
   d) no correlation [1]   e) negative [1]   f) positive [1]
   g) positive [1]
3. A: a) no correlation [1]   b) there is no association between daily rainfall and price [1]
   B: a) negative [1]   b) higher prices are associated with lower sales [1]
   C: a) positive [1]   b) lower rainfall is associated with lower sales [1]

1.

Destination	Number of Girls	Angle
Majorca	300	120°
Orlando	225 [1]	90°
Paris	50	20° [1]
Barbados	325 [1]	130° [1]

2. a)

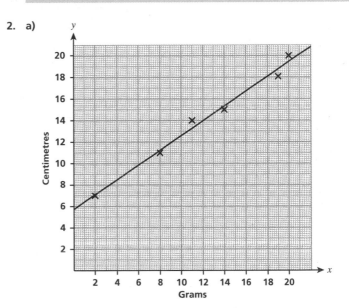

[3 marks for correct plotting of all 6 points; –1 per error]
   b) positive [1]
   c) i) any number between 9cm and 11cm [1]
      ii) any number between 15.5g and 16.5g [1]
      iii) any number between 5cm and 6cm [1]

3. a) $3x + 234 = 360$ [1] has solution, $x = 42$ [2 marks: 1 for accuracy; 1 for method]
   b) The angle for 'sandwiches' is 84° [1] so the total number of students is $\frac{360}{84} \times 266 = 1140$ [1]. The number choosing pizza is $\frac{114}{360} \times 1140 = 361$ [1].

Pages 43–46

Algebra

1. a) $5m$ [1]   b) $2x$ [1]   c) $9t$ [1]   d) $\frac{b}{3}$ [1]   e) a [1]   f) n^2 [1]
2. a) $9d$ [1]   b) $2z$ [1]   c) p [1]   d) $3x$ [1]   e) 0 [1]   f) $4u$ [1]
3. a) $6g + 12$ [1]   b) $7x - 28$ [1]   c) $14 - 2d$ [1]   d) $27 + 9v$ [1]
   e) $6z - 6$ [1]   f) $2a - 30$ [1]
4. a) $6d$ [1]
   b) $5x + 5$ [2 marks: 1 for each term; or 1 for alternative answer $5(x + 1)$]

1. a) $2ab$ [1]   b) $20n$ [1]   c) $\frac{b^2}{2}$ [1]   d) $\frac{2pq}{5}$ [1]   e) $18j$ [1]   f) $10t^2$ [1]
2. a) $d + c$ [1]   b) $1 + 9f$ [1]   c) $6y$ [1]   d) $r^2 + 2r$ [1]   e) $4 + 3w$ [1]
   f) $-4x - 2 + x^2$ [1]

3. a) 20 [2] b) $8p$ [2] c) $6z$ [2] d) $8 + 3a$ [2] e) $8a + 16$ [2]
   f) $16 + 32w$ [2]
4. A: a) $2d + 10$ [2] b) $3d + 6$ [2]
   B: a) $4b + 30$ [2] b) $24b + 36$ [2]
5. a) a^2b^2 [1] b) $6f^4$ [1] c) $6t^4$ [1] d) $3e^6$ [1]

1. A: a) $4x + 10$ [2 marks: 1 for each term]
      b) $2(x + 1) + 2(x + 2)$ [1] which simplifies to $4x + 6$ [1]
   B: a) $6x + 26$ [2 marks: 1 for each term]
      b) $12x + 4(2x + 1)$ [1] which simplifies to $20x + 4$ [1]
2. a) $9x^4$ [2 marks: 1 for 9; 1 for x^4]
   b) $12(x + 5)$ [1] which simplifies to $12x + 60$ [1]
3. The other side has length, $2p + 1$ [2] so the area is $3(2p + 1)$ [1] which simplifies to $6p + 3$ [1]
4. a) They are the same [1]
   b) The side numbers are $\frac{1}{2}(a + b)$, $\frac{1}{2}(b + c)$, $\frac{1}{2}(c + d)$, $\frac{1}{2}(d + a)$ [1]
   sum of the corner numbers = $a + b + c + d$ [1]
   sum of the side numbers $= \frac{1}{2}(a + b) + \frac{1}{2}(b + c) + \frac{1}{2}(c + d) + \frac{1}{2}(d + a)$ [1]
   $= \frac{a}{2} + \frac{b}{2} + \frac{b}{2} + \frac{c}{2} + \frac{c}{2} + \frac{d}{2} + \frac{d}{2} + \frac{a}{2}$ [1]
   $= a + b + c + d$ [1]

Pages 47–50

Congruence and Scaling

1. A and H [1] B and D [1] C and F [1] E and G [1]
2. a) $3 : 7$ [1] b) $3 : 4$ [1] c) $6 : 5$ [1]
3.

Length on scale drawing	Actual length
2cm	8m [1]
5cm	20m [1]
4cm [1]	16m
7cm [1]	28m

1.

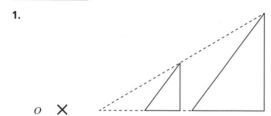

o ✗

[3 marks: 1 for the construction lines; 2 for the triangle]

Measure the sides of the new triangle to check that they are double the length of the original.

2. a)

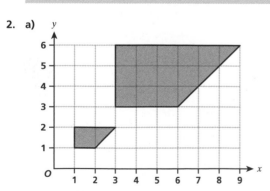

[4 marks: 1 for each corner]

b) i) 1.5cm² [2] ii) 13.5cm² [2]
   c) 1 : 9   [1]
3. a) 1 : 8 [1] b) 20 : 1 [1] c) 3 : 4 [1]
4. a) 1.5km [2 marks; give 1 for 150 000cm]
   b) 20cm [2 marks; give 1 for 0.0002km]

1. a) A and D [2 marks for both letters; 0 otherwise] b) 2.4 [1]

Work systematically through all possible pairs. In each case you need to check if the sides are in the same ratio. For example, when comparing A and B notice that $\frac{7.2}{4}$ is not the same as $\frac{9.2}{5}$ so these are not enlargements.

2. a) $A\hat{C}B = X\hat{Z}Y$ [1] $AC = XZ$ [1] $CB = ZY$ [1] SAS [1]
   b) $RQ = UT$ [1] $PR = SU$ [1] $PQ = ST$ [1] SSS [1]
3. a) 2   [1]
   b) 6cm   [1]
   c) 22 : 44 [1] 1 : 2 [1]
   d) 24 : 96 [1] 1 : 4 [1]
   e) i) m is the scale factor [1] ii) n is the square of the scale factor [1]

Pages 51–53

Fractions and Decimals

1. a) $\frac{1}{16}$ [2] b) $\frac{11}{15}$ [2] c) $\frac{16}{35}$ [2] d) $\frac{13}{24}$ [2] [2 marks: 1 method; 1 accuracy]

Use common sense to check that your final answers seem about right.

2. a) $4\frac{1}{2}$ [1] b) $2\frac{2}{5}$ [1] c) $10\frac{1}{3}$ [1] d) $4\frac{1}{2}$ [1]
3. a) 24 [1] b) 18 [1] c) $\frac{1}{10}$ [1] d) $\frac{1}{21}$ [1]
4. a) 1400 [1] b) 12 000 [1] c) 200 [1] d) 12 [1]

1. a) $5\frac{11}{20}$ [2] b) $2\frac{3}{8}$ [2] c) $11\frac{1}{12}$ [2] d) $2\frac{2}{3}$ [2] [2 marks: 1 method; 1 accuracy]
2. a) 14 [2] b) $16\frac{1}{2}$ [2] c) $1\frac{1}{4}$ [2] d) $\frac{3}{4}$ [2] e) 16 [2] f) $\frac{3}{5}$ [2]
   g) $18\frac{2}{3}$ [2] h) $1\frac{11}{12}$ [2] [2 marks: 1 method; 1 accuracy]
3. a) 0.012 [2] b) 120 [2] c) 4 [2] d) 0.009 [2] e) 90 [2] f) 20 [2]
   g) 6 [2] h) 0.03 [2] [2 marks: 1 method; 1 accuracy]
4. a) 266.4 [1] b) 2664 [1] c) 266.4 [1] d) 72 [1] e) 3.7 [1]
   f) 3700 [1]
5.

X	0.1	4 [1]	30
0.3	0.03 [1]	1.2	9 [1]
20	2 [1]	80 [1]	600 [1]
300 [1]	30 [1]	1200 [1]	9000

1. Area: $9\frac{5}{8}$ cm² [2 marks: 1 method; 1 accuracy]
   Perimeter: $12\frac{1}{2}$ cm [2 marks: 1 method; 1 accuracy]
2. $2 \times 2\frac{2}{5} = 4\frac{4}{5}$ [1] $3 \times 1\frac{3}{4} = 5\frac{1}{4}$ [1] so total cut off is $10\frac{1}{20}$ m [1] with $1\frac{19}{20}$ m remaining [1]
3. a) 2 605 451 [2] b) 25 267.2 [2] c) 12 034 [2] d) 1424.505 [2]
   e) £96 118 [2]
   [2 marks for the correct answer to each part provided it is clear that the results of the given calculations have been used; 0 otherwise e.g. in c) notice that $\frac{2}{17}$ is half of $\frac{4}{17}$ so the answer is worked out by dividing the given value of 24 068 by 2]

Proportion

1. a) £8 [1]   b) £40 [1]
2. a) 16 [1]   b) 48 [1]
3. a) 28hrs [1]   b) 14hrs [1]

> Ask yourself whether a question is on direct or inverse proportion before you begin.

4. a) 18hrs [1]   b) 2hrs [1]

1.

Gallons	2	4	18	6 [1]	14 [1]
Litres	9	18 [1]	81 [1]	27	63

2.

Number of lawnmowers	4	2	16	8 [1]	1 [1]
Time (mins)	56	112 [1]	14 [1]	28	224

> As a check on inverse proportion questions remember that the numbers in each column multiply to give the same value which in this case is 224.

3.

Number of brownies	15	25
Dark chocolate	225g	375g [1]
Eggs	3	5 [1]
Ground almonds	150g	250g [1]

1. a) $y = 25x$ [2]

b)

Length of material x (metres)	1	3	8
Cost, £y	25 [1]	75 [1]	200 [1]

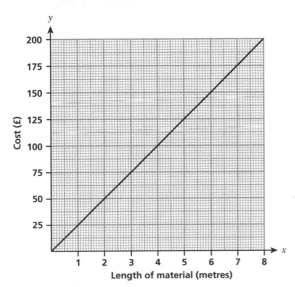

[2 marks for line]

c) 5m   [1]

2. $t = \frac{42}{n}$ [2]
3. a) A [1]   b) C [1]   c) C [1]   d) C [1]   e) B [1]   f) C [1]

Circles

1. a) 14πcm [1]   b) 8πmm [1]   c) 10πm [1]

> Make sure that you know whether you are given the diameter or radius in each part before you begin.

2. a) 49πcm^2 [1]   b) 16πmm^2 [1]   c) 25πm^2 [1]
3. a) [3 marks: 1 mark for each correct label]

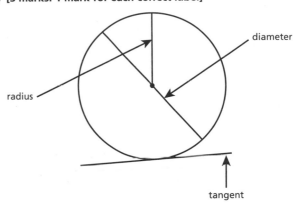

b) [4 marks: 1 mark for each correct label]

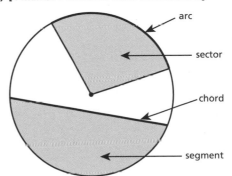

1. A: a) arc length = $\frac{1}{2} \times 2\pi \times 4$ [1] perimeter = $\frac{1}{2} \times 2\pi \times 4 + 8$ [1]
       which is 20.6cm [1]

   b) $\frac{1}{2} \times \pi \times 4^2 = 25.1$ cm^2 [2 marks: 1 method; 1 accuracy]

   B: a) arc length = $\frac{1}{4} \times 2\pi \times 7.5$ [1] perimeter = $\frac{1}{4} \times 2\pi \times 7.5 + 15$
       [1]   which is 26.8cm [1]

   b) $\frac{1}{4} \times \pi \times 7.5^2 = 44.2$cm^2 [2 marks: 1 method; 1 accuracy]

> A common mistake is to forget to add on the lengths of the straight lines when finding the perimeter.

2. a) $2\pi \times 12 = 75.4$m [2 marks: 1 method; 1 accuracy]

   b) $\frac{75.4}{80} \times 60 = 57$s [2 marks: 1 method; 1 accuracy]

3. a) $\pi \times 15^2 = 706.9$ [2 marks: 1 method; 1 accuracy]
   b) circular costs 1.27p per cm^2 [1]   square costs 1.74p per cm^2 [1]
   so circular is best value [1]

1. $C = 251.3$cm [1] $\frac{10000}{251.3} = 39.79$ [1] so 39 **complete** revs [1]
2. a) rectangle = 50×34 [1] circle = $\pi \times 17^2$ [1] add to get 2610cm^2
   [2 marks: 1 method; 1 rounding]
   b) square = 23^2 [1] quarter circle = $\frac{1}{4}\pi \times 15^2$ [1] subtract to get
   352cm^2 [2 marks: 1 method; 1 rounding]
   c) square = 24^2 [1] circle = $\pi \times 12^2$ [1] subtract to get 124cm^2
   [2 marks: 1 method; 1 rounding]

Equations and Formulae

1. a) $x = 8$ [2] b) $x = 3$ [2] c) $x = -2$ [2] d) $x = 10$ [2] [2 marks: 1 method; 1 accuracy. The method mark is awarded for appropriate working e.g. in a) $2x = 16$ must be seen.]
2. a) $x = 3$ [2] b) $x = 5$ [2] c) $x = 3$ [2] d) $x = 4$ [2] [2 marks: 1 method; 1 accuracy. The method mark is awarded for appropriate working e.g. in b) $2x - 3 = 7$ and $2x = 10$ must be seen.]
3. a) $2x + 4 = 14$ [2 marks: 1 for each side of the equation]
    b) $x = 5$ [2 marks: 1 for $2x = 10$; 1 for $x = 5$]

---

1. a) $x = 2$ [2] b) $x = 5$ [2] c) $x = 6$ [2] d) $x = 13$ [2] e) $x = 1$ [2]
    f) $x = 7$ [2] [2 marks: 1 method; 1 accuracy. The method mark is awarded for appropriate working e.g. in d) $5x - 20 = 3x + 6$ and $2x = 26$ must be seen.]

    You can check your final answer by substituting it back into the original equation.

2. a) $g = f + h$ [1]
    b) $d = vt$ [1]
    c) $x = \frac{1}{2}(y - 6)$ [2] [2 marks: 1 method; 1 accuracy]
    d) $b = \frac{1}{3}(d - 2a)$ [2] [2 marks: 1 method; 1 accuracy]
3. $4x + 10 = 8x + 6$ [2 marks: 1 for each side of the equation]
    $x = 1$ [2 marks: 1 method; 1 accuracy]

---

1. $5(x - 2) = 2(4 - x)$ [2 marks: 1 for each side of the equation]
    $5x - 10 = 8 - 2x$ [1]
    $x = \frac{18}{7}$ [2 marks: 1 method; 1 accuracy]
    $2\frac{4}{7}$ [1]
2. $\frac{7}{2}(x + 1) = 2(x + 4)$ [2 marks: 1 for each side of the equation]
    $7x + 7 = 4x + 16$ [1]
    $x = 3$ [2 marks: 1 method; 1 accuracy]
3. a) $\frac{2A}{a + b}$ [2 marks: 1 method; 1 accuracy]
    b) 6 [2]
4. a) $t = \frac{C - b}{a}$ (or equivalent) [2 marks: 1 method; 1 accuracy]
    b) $a = 30$ [1] $b = 50$ [1] $t = 6$ [1]
5. $8x - 18 = 5x + 6$ [1] has solution, $x = 8$
    [2 marks: 1 for method; 1 for accuracy]
    length = 46 [1] width = 20 [1] so the area is 920cm² [1].

Pages 66–69

Comparing Data

1. a) $30 < T \leqslant 45$ [1] b) $15 < T \leqslant 30$ [1] c) $30 < T \leqslant 45$ [1]
    d) 28 [1]

    The symbol $<$ means 'less than' and $\leqslant$ means 'less than or equal to' so the interval $30 < T \leqslant 45$ excludes 30 but includes 45.

2. 4.2 [1] 2 [1] 2 [1] 9 [1]

---

1. a)

Journey times, t (minutes)	Frequency
$0 < t \leqslant 10$	3 [1]
$10 < t \leqslant 20$	6 [1]
$20 < t \leqslant 30$	7 [1]
$30 < t \leqslant 40$	4 [1]

b) i)

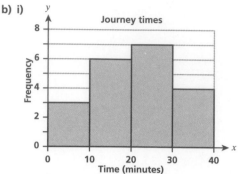

[2 marks: –1 for any error]

ii)

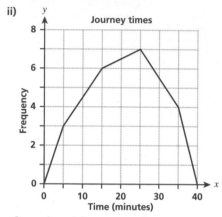

[2 marks; –1 for any error]

c) 45% [1]

2. a) **Mode:** £18 000 [1] **Median:** £27 500 [2]
    Mean: £62 166.67 [2]
    b) Median is best [1] acceptable suitable answer, e.g. Mode is misleading since it gives the salary of the two lowest paid [1]. Mean is too high; it is distorted by the one high salary [1].
3. Rory [1] because his average score is lower [1] and he is the more consistent player since he has a lower range [1].

    Compare both the average and spread.

---

1. 1, 1, 2, 4, 7 [2 marks for all correct numbers; otherwise 0]
2. a) We do not know the figures before they were grouped. [1] There are five days when she sells between zero and four pairs of shoes, and one day when she sells between 24 and 28 pairs. The largest value is therefore $28 - 0 = 28$ [1].

    Spend time making sure that you understand what a diagram represents before you begin.

    b)

[3 marks: –1 per mistake]

c) Hermione's modal class is $16 < s \leqslant 20$ and Celia's is $12 < s \leqslant 16$ [1] so Hermione has the better modal class [1].
d) Celia is better even though her modal class is lower. [1] There are only a few days where her sales are low and she has many days when she sells far more shoes than Hermione. [1]

5 Use a calculator to work out the following.
   Give your answers in standard form rounded to 2 significant figures.

   a) $(7.93 \times 10^2) \times (9.54 \times 10^7) =$ _____

   b) $(8.549 \times 10^3) \times (6.653 \times 10^2) =$ _____

   c) $(3.816 \times 10^9)^2 =$ _____   [3]

   Total Marks _____ / 22

---

1 Without using a calculator work out each of the following.
   Express your answers in standard form.

   a) $(6 \times 10^3) \times (4 \times 10^5) =$ _____   [1]

   b) $(2 \times 10^8) \times (6 \times 10^9) =$ _____   [1]

   c) $(5 \times 10^4) \times (6 \times 10^9) =$ _____   [1]

   d) $(7 \times 10^{10})^2 =$ _____   [1]

   e) $(4 \times 10^6)^3 =$ _____   [1]

(FS) **2** In November 2013 total personal debts in the UK reached £1 429 624 000 000.
(PS)
   a) Write this figure in standard form. _____   [1]

   b) Given that there were about 26.5 million households in the UK, estimate the average
   household debt. Give your answer as an ordinary number, rounded to the nearest thousand.

   _____

   _____   [2]

(PS) **3** One light-year is the distance light travels in one year. A star is estimated to be 56 light years
(MR)   away from the Earth. Use the fact that the speed of light is 3×10^8 m/s to estimate the distance
   of the star in kilometres. Give your answer in standard form rounded to 2 significant figures.

   _____

   _____

   _____   [4]

   Total Marks _____ / 12

_____ / 11

_____ / 22

_____ / 12

How do you feel about these skills?

(PS) (MR) (FS)

Green = Got it!
Orange = Nearly there
Red = Needs practice

Interpreting Data

1 State whether these scatter diagrams display **positive**, **negative** or **no correlation**.

a)

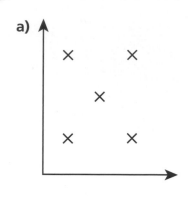

....

b)

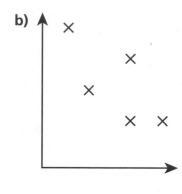

....

c)

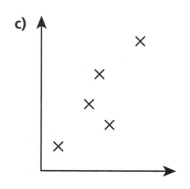

....

[3]

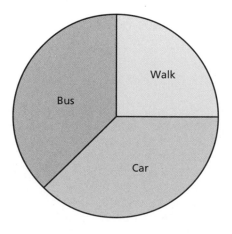

(PS) **2** Twenty-four students in form 8A were asked how they travelled to school. Their responses are shown in the pie chart below.

Walk

Bus

Car

a) How many walked to school? .... [1]

b) How many travelled to school by bus? .... [1]

3 Thirty-two students in form 8B were asked to name their favourite fruit and the results are given in the table below.

Apple	Banana	Orange	Grape
8	16	4	4

Complete the pie chart.

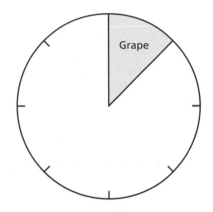

[3]

Total Marks .... / 8

1 A teacher noted the test grades for a group of 72 boys and displayed the results in a pie chart.

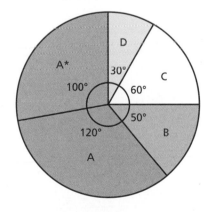

Use the pie chart to complete the table below.

Test Result	Angle	Fraction	Number of boys
D			
C	60°	$\frac{60}{360} = \frac{1}{6}$	$\frac{1}{6} \times 72 = 12$
B			
A			
A*			

[8]

(PS) **2** State whether you would expect there to be **positive**, **negative** or **no correlation** between these variables:

a) student's Spanish and French marks _____ [1]

b) train fare and length of journey _____ [1]

c) maximum outside temperature and number of woolly hats sold _____ [1]

d) student's height and maths mark _____ [1]

e) outside temperature and daily household heating costs _____ [1]

f) person's weight and mean daily calorie consumption _____ [1]

g) age and height of a cactus plant _____ [1]

(PS) **3** A study was made into the daily sales and price of raincoats together with daily rainfall. The following scatter graphs were then plotted.

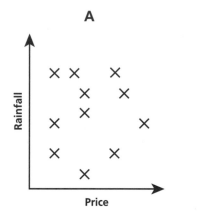

A

B

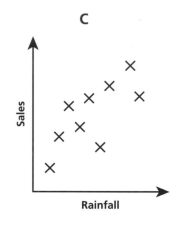

C

In each case:

a) state whether there is **positive**, **negative** or **no correlation**

b) describe in words what this correlation means in each context.

A: **a)** _____ [1]

   b) _____ [1]

B: **a)** _____ [1]

   b) _____ [1]

C: **a)** _____ [1]

   b) _____ [1]

Total Marks _____ / 21

PS
MR

1 A group of girls were asked to choose their favorite holiday destination from a list of four. This data, together with the angle that would be needed to represent this information on a pie chart, is shown in the table. Fill in the missing numbers.

Destination	Number of girls	Angle
Majorca	300	120°
Orlando	....	90°
Paris	50	....
Barbados	....	....

[4]

PS

2 An elastic string is suspended from the ceiling. This is then stretched vertically by hanging weights at the other end. The table shows the masses of the weights (in grams) and the length of the elastic (in centimetres).

x (grams)	8	19	11	14	2	20
y (centimetres)	11	18	14	15	7	20

a) Plot these points on the grid.

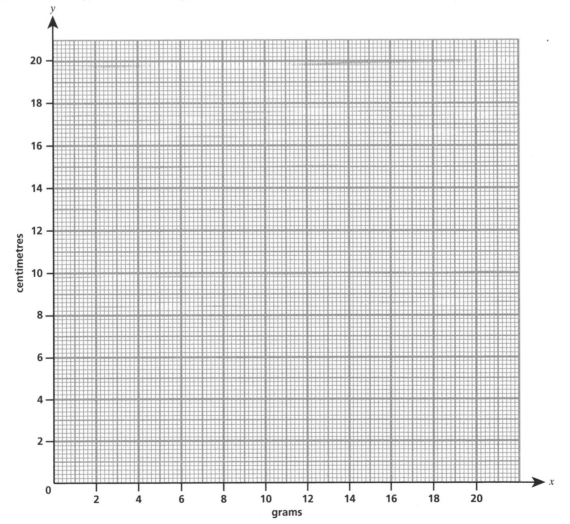

[3]

b) Describe the correlation displayed in the scatter diagram in part **a)**.

.... [1]

c) By drawing a line on the diagram that passes close to all six points, estimate the:

   i) length when the mass is 6g .... [1]

   ii) mass needed to produce a length of 17cm .... [1]

   iii) unstretched length of elastic .... [1]

(PS) (MR) **3** A school canteen offers students a choice of salads, sandwiches, pasta or pizza for lunch one day.
The numbers choosing each option were recorded and the results displayed in the following pie chart.

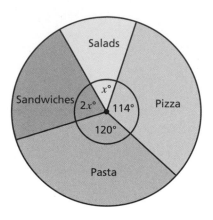

a) Work out the value of x in the pie chart.

....

....

.... [3]

b) If 266 students chose a sandwich, how many chose a pizza?

....

....

.... [3]

Total Marks .... / 17

.... / 8

.... / 21

.... / 17

How do you feel about these skills?

  Green = Got it!
Orange = Nearly there
Red = Needs practice

Algebra

1 Write each of these expressions as simply as possible.

    a) $5 \times m =$ _____

    b) $x + x =$ _____

    c) $t \times 9 =$ _____

    d) $b \div 3 =$ _____

    e) $a \times 1 =$ _____

    f) $n \times n =$ _____   [6]

2 By collecting like terms simplify each expression.

    a) $3d + 6d =$ _____

    b) $10z - 8z =$ _____

    c) $6p - 5p =$ _____

    d) $5x - 3x + x =$ _____

    e) $4x + 2x - 6x =$ _____

    f) $7u - 2u - u =$ _____   [6]

3 Expand the brackets.

    a) $6(g + 2) =$ _____

    b) $7(x - 4) =$ _____

    c) $2(7 - d) =$ _____

    d) $9(3 + v) =$ _____

    e) $6(z - 1) =$ _____

    f) $2(u - 15) =$ _____   [6]

4 Write down simplified expressions for the perimeters of each shape.

    a)

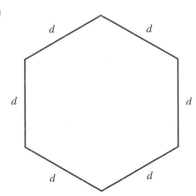

_____ [1]

    b)

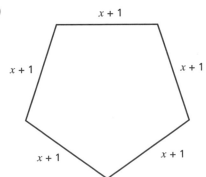

_____ [2]

Total Marks _____ / 21

1. Write each of these expressions as simply as possible.

   a) $a \times 2 \times b =$ _____

   b) $5n \times 4 =$ _____

   c) $b \times \frac{1}{2} \times b =$ _____

   d) $\frac{2}{5} \times p \times q =$ _____

   e) $3 \times 6j =$ _____

   f) $5t \times 2t =$ _____ [6]

2. By collecting like terms simplify each expression.

   a) $4d + c - 3d =$ _____

   b) $2 + 7f - 1 + 2f =$ _____

   c) $3x + 5y - 3x + y =$ _____

   d) $3r^2 + 4r - 2r^2 - 2r =$ _____

   e) $3 + 6w - 3w + 1 =$ _____

   f) $4x - 3 + x^2 - 8x + 1 =$ _____ [6]

3. Expand the brackets and simplify.

   a) $4(g + 5) - 4g =$ _____ [2]

   b) $6(p + 4) + 2(p - 12) =$ _____ [2]

   c) $30 + 2(3z - 15) =$ _____ [2]

   d) $2(4 + a) + a =$ _____ [2]

   e) $6(a + 5) + 2(a - 7) =$ _____ [2]

   f) $2(3 + w) + 5(2 + 6w) =$ _____ [2]

4. Write down simplified expressions for each rectangle for the a) perimeter and b) area.

   A

   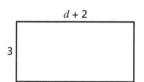

   B

   a) _____

   a) _____

   b) _____

   b) _____ [8]

5. Write each expression in index form.

   a) $a \times b \times b \times a =$ _____

   b) $2f \times f \times f \times 3f =$ _____

   c) $t \times t \times 3t \times t \times 2 =$ _____

   d) $e \times e \times e \times e \times e \times 3e =$ _____ [4]

   Total Marks _____ / 36

MR **1** Write down simplified expressions for each shape for the **a)** perimeter and **b)** area.

A

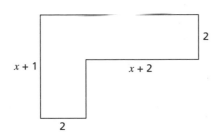

B

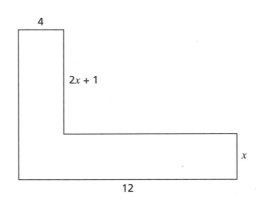

a) ....

a) ....

....

....

....

....

b) ....

b) ....

....

....

....

.... **[8]**

2 Write down a simplified expression for the volume of each cuboid.

a)

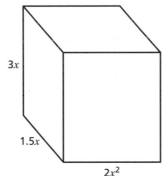

b)

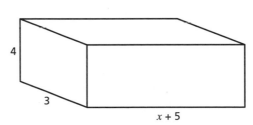

....

....

....

.... **[2]**

MR **3** One side of a rectangle has length, 3, and its perimeter is $4p + 8$.
Find a simplified expression for the area of this rectangle.

....

....

....

.... **[4]**

(MR) **4** Diagram 1 shows a square with the numbers, 3, 7, 13 and 25 written in the corners.
In Diagram 2 the average (that is the mean) of the corner numbers 3 and 7 has been worked
out and the answer, 5, has been written in the middle of the side. The same thing has been
done to the other pairs of corner numbers.

Diagram 1

```
3              7



25            13
```

Diagram 2

```
3      5      7

14            10

25     19     13
```

The sum of the corner numbers is: 3 + 7 + 13 + 25 = 48.

The sum of the side numbers is 5 + 10 + 19 + 14 = 48.

a) What do you notice about these sums? .... [1]

b) Use algebra to show that this always works by completing Diagram 2 below for the case
when general numbers, a, b, c and d are put in the corners.

Diagram 1

```
a              b



d              c
```

Diagram 2

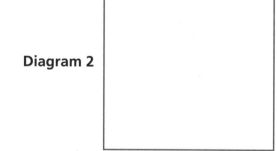

....

....

....

.... [5]

Total Marks .... / 20

.... / 21

.... / 36

.... / 20

Congruence and Scaling

1 Which pairs of rectangles are congruent?

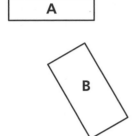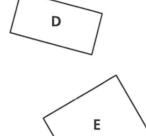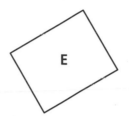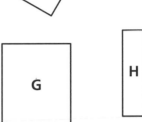

.... and ....

.... and ....

.... and ....

.... and ....   [4]

2 Simplify the following ratios.

a) 15mm : 35mm ....

b) 6m : 8m ....

c) 54cm : 45cm ....   [3]

(PS) **3** Complete the table below for a scale drawing in which the scale is 1cm : 4m.

Length on scale drawing	Actual length
2cm	....
5cm	....
....	16m
....	28m

[4]

1 Enlarge the triangle by scale factor 2 about the centre O.
Leave your construction lines clearly on the diagram.

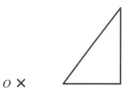

[3]

2 **a)** Enlarge the trapezium by a scale factor of 3 about the origin O. [4]

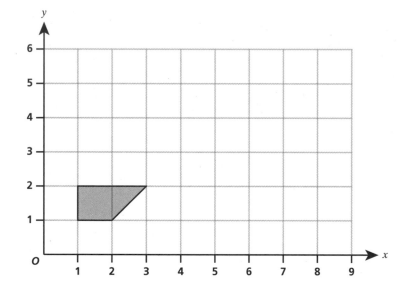

b) If the grid lines are 1cm apart, write down the area of the original trapezium and the area of its enlargement.

   i) Area of original = ... cm² [2]

   ii) Area of enlargement = ... cm² [2]

c) Write down the ratio, simplified as far as possible, for the area of the original trapezium to the area of the enlargement.

.... [1]

3 Express each of the following ratios in its simplest form.

   a) 25cm : 2m   **b)** 18cm : 9mm   **c)** 750m : 1km

[3]

The scale on a map is 1 : 50 000.

a) A road is 3cm long on the map. What is the actual length of the road in kilometres?

....

.... km [2]

b) The distance between two towns is 10km.
How far apart are they on the map in centimetres?

....

.... cm [2]

Total Marks .... / 19

(MR) ① Five rectangles have the dimensions given in the list below. All lengths are in centimetres. Identify the pair that are enlargements of each other and state the scale factor of enlargement.

A: 4 by 5   **B:** 7.2 by 9.2   **C:** 10 by 15   **D:** 9.6 by 12   **E:** 100 by 120

a) The two rectangles are .... and ....   [2]

b) The scale factor of enlargement is ....   [1]

(MR) ② Show that each of the following pairs of triangles are congruent.
Give reasons for each answer and state which condition of congruency you are using:
SSS, SAS or *ASA*.

a)

....

....

....

....   [4]

b)

....

....

....

....   [4]

3 Rectangle B is an enlargement of rectangle A.

A   3cm

8cm

B x

16cm

a) Write down the scale factor of enlargement. .... [1]

b) Write down the value of x. .... [1]

c) Find the ratio of the perimeter of A to the perimeter of B, writing your answer in the
form 1 : m.

....

.... [2]

d) Find the ratio of the area of A to the area of B, writing your answer in the form, 1 : n.

....

.... [2]

e) How are the values of m and n in parts c) and d) related to the scale factor of
enlargement in part a)?

 i) value of m is .... [1]

 ii) value of n is .... [1]

Total Marks .... / 19

.... / 11

.... / 19

.... / 19

Fractions and Decimals

Calculators should **not** be used for any of the questions in this chapter.

---

1 Work out the following. Give your answers in their lowest terms.

   a) $\frac{13}{16} - \frac{3}{4}$ b) $\frac{2}{15} + \frac{3}{5}$ c) $\frac{6}{7} - \frac{2}{5}$ d) $\frac{3}{8} + \frac{1}{6}$

   ....

   .... [8]

2 Work out the following. Give each answer as a mixed number reduced to its lowest terms.

   a) $\frac{1}{4} \times 18$ b) $\frac{1}{5} \times 12$ c) $\frac{1}{3} \times 31$ d) $\frac{1}{6} \times 27$

   ....

   .... [4]

3 Work out.

   a) $6 \div \frac{1}{4}$ b) $9 \div \frac{1}{2}$ c) $\frac{1}{5} \div 2$ d) $\frac{1}{7} \div 3$

   .... [4]

4 Work out.

   a) 20×70 b) 6×2000 c) $4000 \div 20$ d) $3600 \div 300$

   .... [4]

Total Marks _____ / 20

---

1 Work out the following. Give each answer as a mixed number reduced to its lowest terms.

   a) $2\frac{1}{4} + 3\frac{3}{10}$ b) $7\frac{5}{8} - 5\frac{1}{4}$ c) $6\frac{1}{3} + 4\frac{3}{4}$ d) $4\frac{5}{12} - 1\frac{3}{4}$

   ....

   .... [8]

2 Work out the following.

Give each answer as a mixed number reduced to its lowest terms.

a) $8 \times 1\frac{3}{4}$

b) $5\frac{1}{2} \times 3$

c) $2\frac{1}{2} \div 2$

d) $3\frac{3}{4} \div 5$

e) $5 \times 3\frac{1}{5}$

f) $1\frac{4}{5} \div 3$

g) $7 \times 2\frac{2}{3}$

h) $5\frac{3}{4} \div 3$

[16]

3 Work out.

a) 0.06×0.2

b) 0.04×3000

c) $2.4 \div 0.6$

d) $0.36 \div 40$

e) 0.3×300

f) $8 \div 0.4$

g) 2000×0.003

h) $240 \div 8000$

[16]

4 Use the fact that $72 \times 37 = 2664$ to write down the answers to the following.

a) $7.2 \times 37 = $ ....

b) $720 \times 3.7 = $ ....

c) $0.72 \times 370 = $ ....

d) $2664 \div 37 = $ ....

e) $266.4 \div 72 = $ ....

f) $2664 \div 0.72 = $ ....

[6]

5 Fill in the missing numbers in the multiplication table.

X	0.1		30
0.3		1.2	
20			
			9000

[9]

Total Marks .... / 55

1 Work out the area and perimeter of a rectangle that is $2\frac{3}{4}$ cm wide and $3\frac{1}{2}$ cm long.

Area: ....

....

Perimeter: ....

.... **[4]**

2 A roll of electrical wire is 12m in length.
An electrician cuts off two pieces each $2\frac{2}{5}$ m long and three pieces each $1\frac{3}{4}$ m long.

Work out the length of wire remaining.

....

....

....

....

.... **[4]**

3 Use the results of the calculations in the box to work out the answers to each part.

$\frac{4}{17} \times 102\,289 = 24\,068$	6.5% of £102 800 = £6682
$\frac{2\,605\,451}{569} - 4579$	$3.129 \times 345 = 1079.505$
$987 \times 256 = 252\,672$	

a) $4579 \times 569 =$ ....

b) $2.56 \times 9870 =$ ....

c) $\frac{2}{17} \times 102\,289 =$ ....

d) $4.129 \times 345 =$ ....

e) 93.5% of £102 800 = ... **[10]**

Total Marks .... / 18

.... / 20	**How do you feel about these skills?**
.... / 55	Green = Got it!
.... / 18	Orange = Nearly there
	Red = Needs practice

53

Proportion

1 If six tickets to the cinema cost £48, find the cost of:

   a) one ticket _____ [1]

   b) five tickets _____ [1]

2 If two minibuses can transport 32 people, how many people can travel in:

   a) one minibus _____ [1]

   b) three minibuses _____ [1]

3 If it took four people 7 hours to paint a fence, how long would it have taken using:

   a) one person _____ [1]

   b) two people _____ [1]

4 If six diggers can dig the foundations of a house in 3 hours, how long would the job have taken using:

   a) one digger_____ [1]

   b) nine diggers _____ [1]

Total Marks _____ / 8

1 Two gallons is approximately 9 litres.

Complete the table to show the conversion between these units of capacity.

Gallons	2	4	18		
Litres	9			27	63

[4]

2 The time taken to mow a lawn is inversely proportional to the number of lawnmowers used. It takes four lawnmowers 56 minutes to mow a lawn.

Complete the table to show the time taken to mow this lawn with different numbers of mowers.

Number of lawnmowers	4	2	16		
Time (mins)	56			28	224

[4]

3 The ingredients for chocolate brownies include: chocolate, eggs and ground almonds.
The quantities required to make 15 brownies are shown in the table.
Complete the last column to show the quantities needed to make 25 brownies.

Number of brownies	15	25
Dark chocolate	225g	
Eggs	3	
Ground almonds	150g	

[3]

Total Marks _____ / 11

1 It costs £100 to buy 4m of curtain material.

a) Find a formula for the cost, £*y*, of buying *x* metres of this fabric.

_____ [2]

b) Use the formula to complete the table and draw a graph of this relationship.

Length of material *x* (metres)	1	3	8
Cost, £*y*			

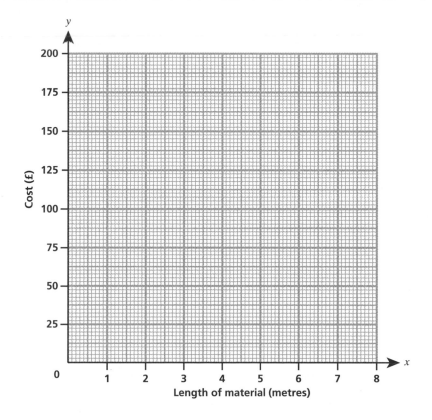

Length of material (metres)

[5]

c) What length of material can be bought for £125? _____ [1]

(MR) **2** It takes six men 7 hours to dig a garden.

Write down a formula for the time, t, taken by n men.

_____ [2]

(MR) **3** For each of the graphs below state whether the variables are:

A: in direct proportion

B: in inverse proportion

C: not in either direct or inverse proportion

a)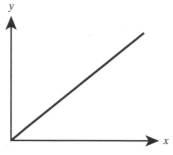

....

b)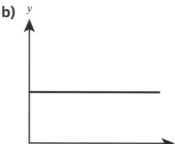

....

c)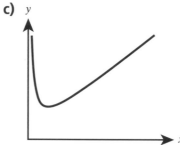

....

d)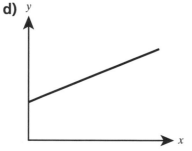

....

e)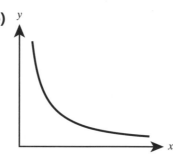

....

f)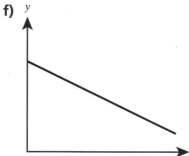

....

[6]

Total Marks .... / 16

.... / 8

.... / 11

.... / 16

Circles

PS **1** Use the formula, $C = 2\pi r$, to calculate the circumference of each circle.
Give each answer as a multiple of π.

a)

b)

c)

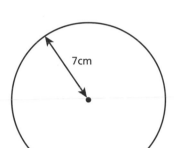

7cm

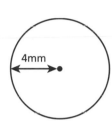

4mm

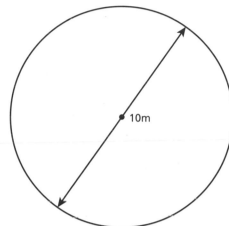

10m

.... [3]

PS **2** Use the formula, $A = \pi r^2$, to calculate the area of each circle in Question 1.
Give each answer as a multiple of π.

a) .... b) .... c) .... [3]

3 a) Write the words **diameter**, **radius** and **tangent** in the spaces on the diagram.

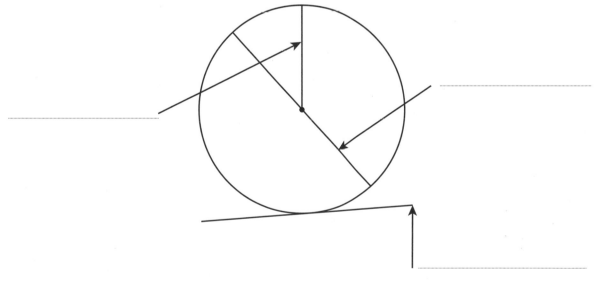

.... [3]

b) Write the words **arc**, **chord**, **segment** and **sector** in the spaces on the diagram.

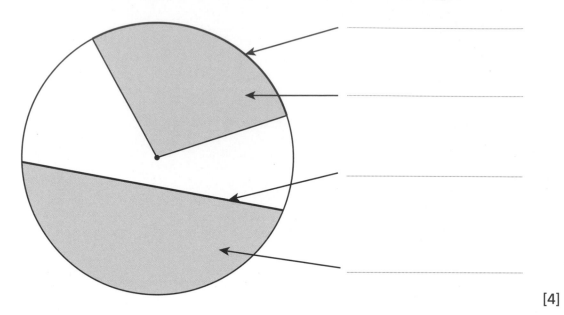

....

....

....

....

[4]

Total Marks .... / 13

(PS) **1** For each of these shapes, find the:
a) perimeter correct to 3 significant figures
b) area correct to 3 significant figures

A

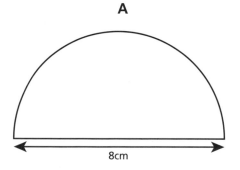

8cm

B

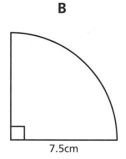

7.5cm

a) ....

....

....

b) ....

....

a) ....

....

....

b) ....

....

[10]

PS **MR** **2** A man walks round the outside of a circular flower bed with a radius of 12m.

a) Work out the circumference of the flower bed to one decimal place.

_____ [2]

b) How long does it take him to walk round the bed if he walks at a constant speed of 80 metres per minute?
Give your answer to the nearest second.

_____

_____ [2]

FS **PS** **MR** **3** Geronimo's pizza is sold in two shapes:

A circular pizza with a radius of 15cm which costs £8.95.

A square pizza with sides 20cm long which costs £6.95.

a) Work out the area of the circular pizza correct to one decimal place.

_____ [2]

b) Which pizza gives the best value for money?
Give reasons for your answer.

_____

_____

_____ [3]

Total Marks _____ / 19

PS **MR** **1** The wheels on a bicycle have a radius of 40cm.

How many **complete** turns does each wheel rotate when the bicycle travels 100m?

_____

_____

_____ [3]

PS
MR

2 Work out the area of each of the shaded regions to 3 significant figures.

a)

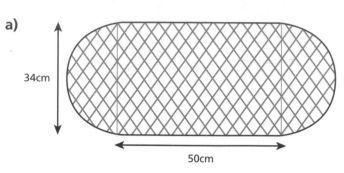

34cm

50cm

Area = ... [4]

b)

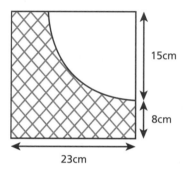

15cm

8cm

23cm

Area = ... [4]

c)

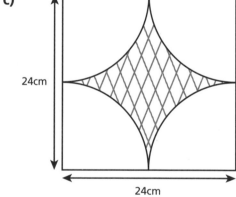

24cm

24cm

Area = ... [4]

Total Marks .... / 15

.... / 13

.... / 19

.... / 15

How do you feel about these skills?

Green = Got it!
Orange = Nearly there
Red = Needs practice

Equations and Formulae

1. Solve these equations.
   Give full details of your working.

   a) $2x - 7 = 9$

   b) $6x + 1 = 19$

   c) $4x + 11 = 3$

   d) $3x - 2 = 28$

   [8]

2. Solve these equations.
   Give full details of your working.

   a) $3x + 1 = 2x + 4$

   b) $6x - 3 = 4x + 7$

   c) $5x + 2 = x + 14$

   d) $7x - 1 = 5x + 7$

   [8]

(PS) 3. The perimeter of the triangle is 14cm.

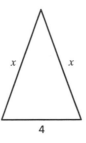

   a) Write down an equation for x.

   [2]

   b) Solve the equation in part a) for x.

   [2]

   Total Marks _____ / 20

1 Solve these equations.

Give full details of your working.

a) $3x + 5 = 7x - 3$

b) $2(x - 1) = 8$

c) $\frac{1}{3}(x + 9) = 5$

d) $5(x - 4) = 3(x + 2)$

e) $6(x + 5) = 9(x + 3)$

f) $\frac{1}{2}(x + 1) = 2x - 10$

[12]

2 Rearrange each of the following formulae to make:

a) g the subject of $f = g - h$

[1]

b) d the subject of $v = \frac{d}{t}$

[1]

c) x the subject of $y = 2x + 6$

[2]

d) b the subject of $d = 2a + 3b$

[2]

(PS) **3** The following shapes have the same perimeter.

Write down an equation for x and solve it.

....

....

....

....

[4]

Total Marks .... / 22

(MR) **1** If I subtract 2 from a number and then multiply the result by 5, I get the same answer when I take the number away from 4 and then double the result.

If my original number is denoted by x, write down an equation and solve it.

Give your answer as a mixed number.

Equation: ....

....

....

....

Mixed number: .... [6]

63

(PS) **2** The triangle and rectangle have the same area.
Write down an equation for x and solve it.

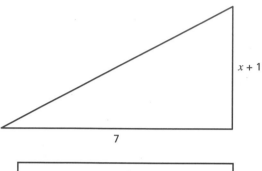

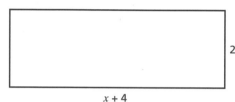

....

....

....

....

....

....

.... [5]

(PS) **3** The formula for the area of a trapezium is $A = \frac{1}{2}(a + b)h$

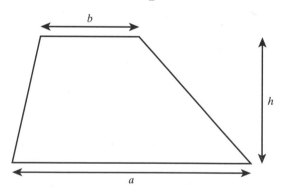

a) Make h the subject of this formula.

....

.... [2]

b) Find the value of h when $a = 4$, $b = 1$ and $A = 15$.

.... [2]

4 The cost, £C, of hiring a party venue for t hours is given by the formula $C = at + b$ where £a is the cost per hour and £b is a fixed booking fee.

a) Make t the subject of the formula.

....

.... [2]

b) It costs £30 an hour to hire the venue and there is an additional fixed booking fee of £50.

If the total cost is £230 use your answer to part a) to work out the length of time that the venue is available.

....

....

.... [3]

5 Find the area of this rectangle.

$8x - 18$

$2x + 4$

$5x + 6$

....

....

....

....

....

Area = ... [6]

Total Marks .... / 26

.... / 20

.... / 22

.... / 26

How do you feel about these skills?

(PS) (MR) (FS)

Green = Got it!
Orange = Nearly there
Red = Needs practice

Comparing Data

(PS) **1** On a recent visit to a theme park, form 8C were asked to make a note of how long they spent in the queue for the Gemini roller coaster.
The results are displayed in the table.

Time, T (minutes)	Frequency
$0 < T \leqslant 15$	4
$15 < T \leqslant 30$	7
$30 < T \leqslant 45$	15
$45 < T \leqslant 60$	2

a) Which is the modal class? .... [1]

b) Jeremy queued for 26mins. In which group was his time recorded? .... [1]

c) Maya queued for 45mins. In which group was her time recorded? .... [1]

d) How many children from form 8C went on the roller coaster? .... [1]

(PS) **2** Work out the **mean**, **median**, **mode** and **range** for these five numbers.

1   2   2   6   10

Mean = ... [1] Median = ... [1] Mode = ... [1] Range = ... [1]

Total Marks .... / 8

(PS) **1** The time taken, in minutes, for a group of 20 children to travel to school is as follows:

18, 5, 28, 22, 12, 9, 23, 31, 17, 18, 28, 25, 31, 20, 36, 39, 30, 8, 13, 22

a) Complete the grouped frequency table.

Journey times, t (minutes)	Frequency
$0 < t \leqslant 10$	....
$10 < t \leqslant 20$	....
$20 < t \leqslant 30$	....
$30 < t \leqslant 40$	....

[4]

b) Illustrate this data by a frequency diagram in the form of a:

   i) block graph

   ii) frequency polygon

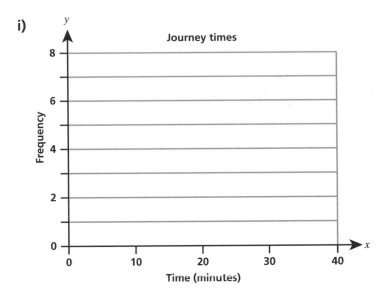

[2]

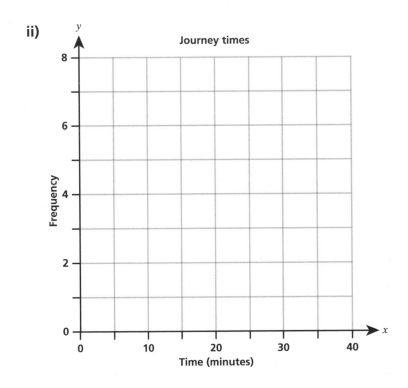

[2]

c) What percentage of these children take 20 minutes or less to travel to school?

[1]

(FS)
(PS)
(MR)

2 A company employs six people and their annual salaries are:

| £18 000 | £18 000 | £25 000 | £30 000 | £32 000 | £250 000 |

a) Find the **mode**, **median** and **mean** salary.
Round the mean to the nearest penny.

Mode = ... [1]

Median = ... [2]

Mean = ... [2]

b) Which of these provides the best measure of the average salary?
Give reasons why the other two averages might be misleading in this case.

....

....

.... [3]

(MR)

3 The table shows the mean and range of golf scores for Rory and Jack after many rounds on a particular golf course.

	Mean	Range
Rory	71	13
Jack	93	25

Who is the better player on this course?
Give reasons for your answer.

....

....

.... [3]

Total Marks .... / 20

(MR)

1 Write down five whole numbers with mode 1, median 2, mean 3 and a range of 6.

.... [2]

2 Hermione works as a sales assistant in a shoe shop.

The number of pairs of shoes that she sells over 35 days is displayed in the frequency diagram.

Celia also works in the same shop.
Her sales figures for the same period are shown in the grouped frequency table.

Daily Sales	Frequency
$0 < s \leq 4$	1
$4 < s \leq 8$	1
$8 < s \leq 12$	3
$12 < s \leq 16$	8
$16 < s \leq 20$	6
$20 < s \leq 24$	3
$24 < s \leq 28$	7
$28 < s \leq 32$	6

a) Explain why it is impossible to work out the exact range of Hermione's sales from this diagram and state the largest value it could be.

_____

_____ [2]

b) Draw a frequency diagram for Celia's sales on the same diagram as Hermione's sales above. [3]

c) Compare the modal classes of these two assistants.

_____ [2]

d) Who is the better sales person? Give a reason for your answer.

_____

_____ [2]

Total Marks _____ / 11

_____ / 8

_____ / 20

_____ / 11

Notes